Essential Physics

Essential Physics

L Kirkup BSc
A J Murkett BSc, PhD
W C Roe BSc, PhD
S D Veazey BSc, MA, CertEd

Department of Science
Luton College of Higher Education

Pitman

PITMAN BOOKS LIMITED
128 Long Acre London WC2E 9AN

Associated Companies
Pitman Publishing New Zealand Ltd, Wellington
Pitman Publishing Pty Ltd, Melbourne

© L Kirkup, A J Murkett, W C Roe, S D Veazey 1983

First published in Great Britain 1983

ISBN 0 273 01896 5

Filmset and printed in Northern Ireland at The Universities Press (Belfast) Ltd.,

Contents

Preface

The appearance of a new textbook at any level poses the inevitable questions: "Why do we need yet another textbook?" and "What parts does it reach that the others conspicuously fail to reach?" By way of reply we would suggest that this book sets out to fill a gap which, in our experience as teachers, is not being filled. The book is aimed in the first place at post O-level and post CSE students who are studying TEC Certificate and Diploma courses in Science particularly and also in Engineering in Colleges of Further or Higher Education, and secondly at first-year sixth-form students who may be unsure about proceeding to A-level or CEE in physics. We hope too that the book may prove to be of value in a wider market and that it will have an appeal to teachers and students at a comparable level in countries outside Great Britain. It is envisaged that students using this book will be following a comparable post O-level course in mathematics. Thus, while we have tried to keep the mathematical development of physics down to a minimum, where necessary we have used differential calculus to deal with rates of change. Such supportive work in mathematics as may be needed can be found in *Electrical and Engineering Mathematics*, volume 2, by R. Meadows, published by Pitman Books.

It should be stressed that this book is not envisaged as an alternative to any of the already established textbooks which are clearly and specifically dedicated to the needs of A-level students who intend to continue with the academic study of physics beyond that level. Our aim is simply to consolidate the principles and essentials of physics without which the more technically minded students cannot apply their knowledge and skills, and equally without which the more academic students cannot proceed with confidence.

We have used SI units and symbols consistently throughout the book, but would not claim to have been unduly rigorous and pedantic in their application. We are firmly of the opinion that physics is only learned satisfactorily through doing, hence there is a liberal use of worked examples in the text and of problems at the end of each chapter. The additional use of illustrative practical work and of further problem solving is strongly recommended to students and teachers using this book. The end-of-chapter problems have been drawn largely from our work with student technicians.

In conclusion, we extend our sincere thanks to our colleagues at Luton, Lee Bonham and Peter Dean, whose comments and criticisms have been liberal, generous and invariably helpful. It is a pleasure to thank John Cushion of Pitman Books who has led us gently into the mysteries of authorship while, at the same time, keeping us firmly on the straight and narrow dictated by deadlines in time and constraints of content. Equally we would wish to record our gratitude to our wives, families and loved ones for their tolerance, forbearance and support during the period when we were struggling to put pen to paper. We also wish to thank numerous students who have worked through most of the problems used and whose difficulties and perceptive questions have caused us to re-consider and to re-evaluate our own understanding of the subject matter. They have all, in their various ways, contributed to what we hope is a substantial reduction in the number of inaccuracies and obscurities, and for which we are very grateful.

(Since the writing of this book, one of the authors, L. Kirkup, has moved to Paisley College of Technology.)

LK
AJM
WCR
SDV
Luton
May 1983

Acknowledgements   The authors gratefully acknowledge the assistance of those who have given permission for the reproduction of photographs:

Fig. 7:19 G Wright, Assistant Director, North East London Polytechnic

Fig. 20:4 From Blackett P. M. S. & Lees D. S., *Proc. Roy. Soc. A* 136, 325, 1932. With permission of the Royal Society.

Fig. 22:6 United Kingdom Atomic Energy Authority.

Some Useful Physical Constants and Values

Universal constant of gravitation $G = 6.67 \times 10^{-11}$ N m^2/kg^2

Acceleration g due to gravity = 9.8 m/s^2

1 radian = 57.3° (π rad = 180°)

Speed of electromagnetic waves in vacuum $c = 3.0 \times 10^8$ m/s

Speed of sound (in air at 20°C) = 340 m/s

Standard atmospheric pressure (1 atmosphere) = 1.013×10^5 Pa
= 760 mmHg

S.T.P. (standard temperature and pressure) $\equiv$ 0°C (273 K), 1 atm.

Permeability of free space $\mu_0 = 4\pi \times 10^{-7}$ H/m

Permittivity of free space $\epsilon_0 = 8.85 \times 10^{-12}$ F/m

Specific heat capacity of water (at 20°C) = 4180 J/kg K

Young's modulus for steel = 2.0×10^{11} N/m^2

Thermal conductivity of copper (at 20°C) = 385 W/m K

Thermal conductivity of air (at 20°C) = 25×10^{-3} W/m K

Mass of one mole of hydrogen = 2×10^{-3} kg

Molar (universal) gas constant $R = 8.31$ J/mol K

Density of mercury = 13 600 kg/m^3 (at S.T.P.)

Density of oxygen = 1.43 kg/m^3 (at S.T.P.)

Density of water = 1000 kg/m^3 (at S.T.P.)

Avogadro's number $N_A = 6.02 \times 10^{23}$ per mole

Surface tension of pure water at 20°C = 72×10^{-3} N/m

Resistivity of copper (at 20°C) = 1.7×10^{-8} Ωm

Temperature coefficient of resistance of copper = 4×10^{-3} per °C

Planck's constant $h = 6.63 \times 10^{-34}$ J s

Mass of electron = 9.1×10^{-31} kg

Charge on electron $e = 1.6 \times 10^{-19}$ C

1 electron-volt (eV) = 1.6×10^{-19} J

1 Ångström = 10^{-10} m

Rydberg constant for hydrogen $R = 1.097 \times 10^7$ per metre

Mass of proton = $1.672\ 614 \times 10^{-27}$ kg = 1.007 277 u

Mass of neutron = $1.674\ 920 \times 10^{-27}$ kg = 1.008 665 u

Mass of neutral hydrogen atom = 1.007 825 u

Unified atomic mass unit $u = 1.66 \times 10^{-27}$ kg

1 Curie = 3.7×10^{10} Becquerel (disintegrations per sec)

Common Physical Quantities and their Units

Displacement (distance) metre   m

Time second   s   (or sec)

Velocity (speed) m/s

Acceleration m/s^2

Mass kilogramme kg

Force newton   N
$$1\,N = 1\,kg\,m/s^2$$

Angular velocity rad/s

Angular acceleration rad/s^2

Torque (moment) newton metre   N m

Moment of inertia kg m^2

Momentum newton second   N s

Impulse newton second   N s

Work joule   J
$$1\,J = 1\,N\,m$$

Energy joule   J

Power watt   W
$$1\,W = 1\,J/s$$

Density kg/m^3

Frequency hertz   Hz
$$1\,Hz = 1\,cycle\ (vibration)/sec$$

Wavelength metre   m

Stress N/m^2

Temperature kelvin   K
$$(°C = K - 273)$$

Specific heat capacity J/kg K

Molar heat capacity J/mol.K

Thermal capacity J/K

Specific latent heat J/kg

Thermal conductivity W/m K

Pressure pascal Pa
$$1\,Pa = 1\,N/m^2$$

Surface tension N/m

Viscosity poiseuille   P
$$1\,P = 1\,kg/m\,s$$

Electric charge coulomb   C
$$1\,C = 1\ ampere\ sec$$

Electric potential volt   V
$$[1\,V = 1\ joule/coulomb]$$

Electric field strength volt/metre   V/m

Capacitance farad   F
$$1\,F = 1\ volt/coulomb$$

Permittivity farad/metre   F/m

Electric current ampere   A
Electrical resistance ohm Ω
Resistivity ohm metre Ω m
Electromotive force (e.m.f.) volt   V
Magnetic field tesla   T
  (Magnetic flux density) $1\ T = 1\ Wb/m^2$
Magnetic flux weber   Wb
Inductance henry   H
Permeability henry/metre   H/m

Dimensionless quantities
Refractive index
Strain
Magnification

1 Introduction

There is no doubt that "we are living in an age of high technology". Behind this statement lies the idea of a world increasingly dependent on the application of science, and in particular the application of physics. Many aspects of engineering and materials science are developments based on the application of physical principles. In the light of this the authors aim to present in the following pages an account of the basics of physics necessary for an up-to-date understanding of the physical world.

At the outset, the question "what is physics?" has to be faced. A glance at the contents page of this book will give some idea of the range of topics included in the subject. The list might also suggest that physics is simply a rather loose grouping of separate and apparently disconnected topics. It might well be asked what the link is between the motion of a satellite in orbit round the earth, the focusing effect of a lens, and the magnetic field associated with an electric current. These three examples illustrate the diversity of phenomena in the physical world which attract the interest of physicists.

While such phenomena are studied with a view to describing and understanding them in their own right, it is essential to appreciate the relationships between phenomena and also to make use of the understanding gained in one area in order to propose explanations in another. The aim therefore is to find an underlying unity between physical processes which are believed to lie at the root of the physical world.

This underlying unity becomes evident, in part, from the fact that in describing the physical world we make use of a number of basic ideas, or concepts, which are useful in a range of different situations. Examples of such ideas which will be found occurring frequently throughout the book include: force, energy, atoms, electrons, fields and waves. It is because these concepts are useful in a number of different areas of physics that the subject has an overall unity.

Returning to the three phenomena mentioned previously we can now see a common feature. All three relate to motion. Satellite motion can be described in terms of the velocity and acceleration of a body moving in the earth's gravitational field; the focusing effect of a lens results from the fact that a beam of light travelling in air enters a different medium (the glass of the lens), in which the velocity has a different value, and undergoes a change of direction; finally, the magnetic field set up round a current-carrying conductor is due to the movement of electrons in the conductor and the size of the field is related to the drift velocity of these electrons. Thus these three separate effects may all be seen to be aspects of motion and its consequences.

In the attempt to put forward an explanation of the physical world frequent reference will be made to *principles, theories* and *laws.*

Principles, like the conservation of momentum and the conservation of energy, are seen as fundamental statements of wide application and are assumed to be always correct. **Theories**, such as the kinetic theory of gases, the electromagnetic theory or atomic theory, are put forward rather more tentatively to explain observations. They are believed, not so much for their absolute correctness, but rather for their usefulness in that they are able to account reasonably well for observations and they also propose a possible mechanism for processes. A good theory also enables its user to make predictions which can be put to the test by experiment. Many theories use models which are really ways of picturing a situation. The kinetic theory is based on the molecular model of matter, while the atomic theory developed at the beginning of this century used Bohr's model of the atom with a nucleus surrounded by electrons moving in certain set circular paths around it. Such models may not be a precise representation of reality, but they are retained nevertheless for their obvious usefulness. Finally, the **laws** of physics express and summarise the observed relationships which are revealed through experimental studies. Examples include Newton's laws of motion (relating motion and force), Coulomb's law (relating force and electric charge), and Ohm's law (relating current and potential difference). Scientific laws are man-made and come about as a result of scientists analysing and evaluating observations, experiments and theories; the physical world is not compelled to obey these laws. They describe the way things are, rather than prescribe the way things should or must be.

The interrelated nature of physics makes it inevitable that some ideas will be met and used before they have been discussed in detail. So, for example, the reader will find that ideas about atomic structure are referred to in chapter 9 as a preliminary to the discussion of the mechanical properties of matter and, again, in distinguishing between conductors and insulators in chapter 13. In such cases cross-references have been included, indicating where a fuller treatment is located.

In conclusion, one or two words of warning which bring the reader back to the ideas set out at the start of this introduction. It would be wrong to limit physics by suggesting that it simply offers a systematic picture of the way in which the world behaves. Its importance for anyone involved in, or planning to start, a science or engineering based career lies in its wider application. Physics is the foundational science and so a basic knowledge of the subject is desirable for the study and understanding of the other scientific subjects. For example, the basic concepts like forces and energy, waves and atoms are essential to the descriptive framework of the biological and earth sciences. Physics finds further application in a wide range of instrumentation and measurement techniques. It is exactly at this point that the overlap with technology occurs. Technology may be described as "the art of finding better ways of doing things" and in order to do this it must make use of the best and most up-to-date tools that so-called "pure" science puts at its disposal. Physics provides many of these tools.

There are, however, certain limitations which affect physics and all the other sciences. It is important to realise that science is not something which is complete, fixed and unchanging. As our knowledge and understanding progress, so our views about laws and theories may be modified or even radically changed. Science and technology cannot provide the definitive answers to all the problems of the 20th and 21st centuries; they can at best only provide means for possible solutions. Moreover, neither science nor technology provides us with the means to decide which problems should take precedence over others.

2 Forces and Motion

Throughout physics you will find frequent reference to forces and motion which indicates that these concepts are basic to the subject. In this chapter we will first outline how motion can be described, then look at the relationship between force and motion in a straight line, and finally we shall look at motion round a circular path.

2:1 Equations of Motion

In linear motion we consider some general object, which is called a **body**, which moves in a straight line. If the body travels distance s in a time t then the average velocity v is given by

$$v = s/t \tag{2:1}$$

In other words, velocity is the rate of moving and its basic unit is metres/sec or m/s. The distance travelled in a straight line, that is in a particular direction, is called the **displacement**. If the velocity does not change in successive time intervals, the motion is called **uniform velocity** motion and could be represented by a graph such as fig. 2:1, where displacement is plotted against time. The slope of this graph (the ratio AB/BC) gives the velocity. Equation (2:1) is the equation, and fig. 2:1 the graph, for uniform velocity motion.

A different motion is shown in fig. 2:2, where the displacement/time curve is non-linear. Here the velocity changes with time and we can then only talk about the velocity at some particular time (which is called the **instantaneous velocity**). At time t_1 for example, the instantaneous velocity is given by the slope at P and may be calculated by drawing the tangent to the curve and determining the ratio AB/BC as before. Mathematically the velocity is written, using the calculus notation, as

$$v = \frac{ds}{dt}$$

This differential of displacement s with time t is equivalent to the slope of the displacement/time graph.

In the case shown in fig. 2:2 the velocity increases with time and the body is then said to accelerate. **Acceleration** is defined as the rate of change of velocity. In many situations we will meet, the acceleration is constant and a number of equations to describe such motion can be derived.

Suppose the velocity of a body changes from an initial value u to a final value v during a time interval t. Then the acceleration a will be

$$a = \frac{v - u}{t}$$

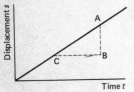

Fig. 2:1 Displacement/time graph for uniform velocity motion. The ratio AB/BC is the velocity.

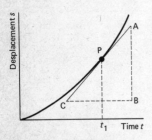

Fig. 2:2 Displacement/time graph for accelerating motion. The slope of the tangent at P gives the instantaneous velocity.

Since this gives the change in velocity per unit time the basic unit for acceleration will be m/s^2. Rearranging this equation gives

$$at = v - u$$

or $v = u + at$ (2:2)

For uniform acceleration, the displacement/time graph would be similar to fig. 2:2 but we could also draw a graph of velocity against time. If the velocity starts at u and increases by equal amounts in successive time intervals, the graph will look like fig. 2:3 and the slope gives the acceleration. The velocity/time graph contains further information because the *area* under the curve, which is shaded in fig. 2:3, gives the total displacement during the motion.

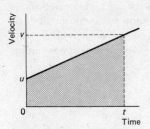

Fig. 2:3 Velocity/time graph for uniform acceleration. The shaded area is the total displacement.

Over time t, under uniform acceleration, the average velocity of the body is $(v + u)/2$ and hence the total displacement is

$$s = \left(\frac{v + u}{2}\right)t$$

If this equation is combined with equation (2:2) we can derive two further relationships. Eliminating t gives

$$s = \left(\frac{v + u}{2}\right)\left(\frac{v - u}{a}\right)$$

$$2as = (v + u)(v - u) = v^2 - u^2$$

hence $v^2 = u^2 + 2as$ (2:3)

If v is eliminated, then

$$s = \left(\frac{u + u + at}{2}\right)t$$

hence $s = ut + \tfrac{1}{2}at^2$ (2:4)

Equations (2:2), (2:3), (2:4) are the **basic equations for motion under uniform acceleration**. It should be noted that these equations cannot be applied to cases where the acceleration changes with time. However these more general motions can still be represented on displacement/time or velocity/time graphs. Non-uniform acceleration occurs in simple harmonic motion which will be discussed in chapter 5.

In any general motion the **instantaneous acceleration** can be found from the slope of a velocity/time graph and mathematically is written as

$$a = \frac{dv}{dt}$$

It is not, however, necessary to be familiar with calculus to be able to handle the problems at the end of this chapter.

Example 2:1

A tennis ball is hit vertically upward and its velocity halves in a time of 0.5 sec. Determine the initial velocity, the maximum height reached and the time the ball is airborne. Any effects due to air resistance may be ignored and the acceleration due to gravity, or acceleration of free fall, may be taken as 9.8 m/s².

Solution Once the ball leaves the bat it moves freely under gravity. Since the **gravitational force** tends to pull the ball towards the earth, its upward velocity will decrease. The upward motion is therefore a retardation (or negative acceleration). In the absence of air resistance, the acceleration is uniform so equations (2:2) to (2:4) can be used as appropriate.

If the initial velocity is u it will decrease to $u/2$ in 0.5 sec. So using equation (2:2),

$$v = u + at \qquad \therefore u/2 = u - (9.8 \times 0.5)$$
$$9.8 \times 0.5 = u - u/2 \quad \text{giving} \quad u = 9.8 \text{ m/s}$$

(Note that the acceleration is negative here.)
At the top of the path the velocity is zero, so from equation (2:3),

$$v^2 = u^2 + 2as \qquad \therefore 0^2 = 9.8^2 - (2 \times 9.8s)$$
$$19.6s = 96.04 \quad \text{giving} \quad s = 4.9 \text{ m}$$

The time to reach the maximum height may be obtained by substituting in equation (2:2),

$$0 = 9.8 - 9.8t \qquad \therefore t = 1 \text{ s}$$

Since the ball accelerates down from the top, at the same rate as it slowed while rising, its total time in the air will be twice this, and hence the ball is airborne for 2 sec.

Example 2:2

Sketch *a*) the velocity/time graph and *b*) the displacement/time graph for the upward part of the motion in Example 2:1.

a) The ball slows down at constant retardation and, since the acceleration is negative, the graph will have negative slope as shown in fig. 2:4. Note that the triangular region under the line has area

$$\tfrac{1}{2} \times 9.8 \times 1$$

which gives 4.9 m as the maximum height.
b) The ball rises to a height of 4.9 m with its velocity decreasing as it rises. The displacement/time graph will therefore have decreasing slope as shown in fig. 2:5. Equation (2:4) can be used to show that after 0.5 sec the ball has risen 3.675 m since

$$s = 9.8 \times 0.5 - \tfrac{1}{2} \times 9.8 \times (0.5)^2 = 4.9 - 1.225 = 3.675 \text{ m}$$

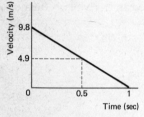

Fig. 2:4 Velocity/time graph for example 2:2.

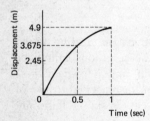

Fig. 2:5 Displacement/time curve for example 2:2.

2:2 The Idea of Force

So far we have only described motion in terms of velocity and acceleration. We now need to ask how motion may be produced in the first place or how the velocity of a body may be changed. The answer which physicists give is that such changes are the result of the action of a **force**. In the example above it was the gravitational force which slowed the ball down and pulled it back to earth.

The relationship between force and motion is summarised in Newton's three laws of motion. **Newton's first law of motion** states that

A body will continue at rest, or move with uniform motion in a straight line, unless a resultant force acts on it.

The law really says three things:

1   A state of rest or uniform velocity implies the absence of any net force.

2   Motion in the absence of forces is linear.

3   A **resultant force** is necessary to produce a change of velocity. The term "resultant force" means that, where more than one force acts on a body, the resulting motion depends on the combined effect. Addition of forces will be dealt with in section 2.5.

The idea that a body could continue indefinitely in motion without the intervention of a force appears to have originated with Galileo. Before his time it was generally believed that motion always required some action to keep the body moving. What Galileo and Newton realised was that a force is involved not in maintaining but in changing the motion of a body. Bodies possess a property called **inertia** which means that their motion will continue unchanged unless and until some force acts. This inertia is measured as the **mass** of a body.

Newton's law sees force as the cause of a change in the *size* of the velocity. A force is also necessary to produce a change in the *direction* of motion since, in the absence of force, motion is in a straight line. If this is the case, which way must a force act to produce a circular path? Consider the case of an orbiting satellite shown in fig. 2:6. Path A represents the trajectory in the absence of a resultant force. To maintain the satellite S in a circular orbit B round the earth, a force *F* must act inwards towards the earth. In this particular case the force is the same gravitational force which influenced the motion of the ball in example 2:1. We shall return to consider the forces in circular motion in section 2:7.

The first law says that the motion of a body will change when a resultant force acts on it. Consider the situation shown in fig. 2:7 where a mass is supported by a spring balance. The body experiences two forces. The earth exerts a gravitational force *F* on the body and there is also a tension force. When the body is hung on the end of the spring, the spring stretches and a force arises within the spring. The spring is said to be in a state of tension and it will exert a force *T*, called the *tension force*, on the body. Although the body experiences two forces, no motion results because they cancel out and produce a resultant force of zero. We will consider other examples where combinations of forces result in a static situation in chapter 3.

Where there are *frictional forces* present, these will act to oppose the motion and the direction of friction is therefore opposite to the direction of the velocity. When a body is accelerating under an applied force, the resultant force on the body will be the difference between the applied force and the frictional force.

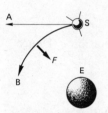

Fig. 2:6 Force required to keep a satellite S in orbit round the earth E.

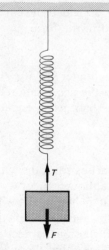

Fig. 2:7 Forces acting on a body suspended from a spring balance.

2:3 Force and Acceleration

Newton's second law of motion states that

The rate of change of momentum of a moving body is proportional to the resultant force on the body.

In effect this quantifies the first law. A moving body has a **momentum** p which is the product of mass and velocity, i.e.

$$p = mv$$

and the action of a resultant force is to change this momentum at a particular rate. (Newton also added that this change takes place along the direction in which the force acts.)

Mathematically the law may be written as

$$F \propto \frac{dp}{dt}$$

which is not the most helpful form for most practical problems. We can however rearrange this relationship, first by assuming that the mass does not change with time, and secondly by a suitable choice of unit for force. If we make the first assumption, the law becomes

$$F \propto m\frac{dv}{dt}$$

The last term in the equation we recognise as the acceleration a. Hence

$$F \propto ma$$

or $\quad F = kma$

where k is some, as yet unspecified, constant. Mass is one of the fundamental quantities in the SI system of units and is measured in kilograms (kg), while acceleration is measured in m/s². If we define our unit of force so that one unit of force accelerates a mass of 1 kg at a rate of 1 m/s², the value of k is unity and we no longer need retain it in the equation. This unit of force is called the NEWTON. Provided the correct units are used, Newton's second law can be applied in the form

$$F = ma \qquad\qquad\qquad (2:5)$$

In example 2:1 above we considered a ball moving under gravity. As it fell from its maximum height the ball accelerated at the acceleration due to gravity (usually denoted by g). The force on the ball, using equation (2:5), is

$$F = mg$$

This force is known as the **weight** of the body and arises from the action of gravity. Notice that weight is strictly a force, although in everyday usage we may talk about a "weight" and quote it in mass units. The process of weighing an object is really a matter of balancing the gravitational forces on each side of a balance. If the gravitational forces are equal then, since g is the same for both sides, it follows that the masses are also equal, and so an unknown mass may be determined.

As well as setting out three laws of motion, Newton also proposed a

Fig. 2:8 Gravitational attraction between two bodies. Distance r is measured between the centres of the bodies.

law of universal gravitation. Any two bodies, of masses m_A and m_B, will exert an attractive force on each other as shown in fig. 2:8. The size of this force depends on the masses and the distance r between the bodies. The **gravitational force equation** is

$$F \propto \frac{m_A m_B}{r^2}$$

or $\quad F = \dfrac{G m_A m_B}{r^2}$

where G is the *universal constant of gravitation* $(6.67 \times 10^{-11}\,\text{N m}^2/\text{kg}^2)$. This force therefore decreases as the masses move further apart.

A body near the earth's surface experiences this gravitational force attracting it towards the earth, so that if the body is free to move it will fall to the ground. This gravitational force is the weight, and for a body of mass m near the earth's surface we can write

$$\frac{G M_E m}{R_E^2} = mg$$

where M_E is the mass and R_E the radius of the earth. It follows from this equation that, if the body is moved some distance from the earth, the gravitational attraction decreases, g also decreases, and the weight of the body is less. The mass however remains unchanged.

2:4 Action and Reaction

The **third law** which Newton put forward may be expressed in the form:
If body A exerts a force (the action) on body B, then body B exerts an equal and opposite force (the reaction) on body A.
Behind this law lies the fact that forces act in pairs and that a force represents an interaction between two bodies. If you had the unfortunate experience of being punched on the chin, your assailant may bruise your jaw but in return crack his (or her) own knuckles. The action bruises your jaw, the reaction (the force exerted by your jaw on him) damages his knuckles. Notice that, as this statement of the law makes clear, the two forces, action and reaction, do not act on the same body. Rather, each interacting body experiences one force.

Consider some more serious examples. Fig. 2:7 shows the forces on a body supported by a spring balance. T as shown is the force exerted by the spring on the mass to hold it. The mass exerts an equal and opposite force on the spring which is responsible for the stretching of the spring.

The gravitational interaction shown in fig. 2:8 is a mutual attraction. The two masses A and B each experience a force attracting them towards the other. Equal and opposite forces F act (one on each body) and these two forces constitute the action and the reaction.

In dealing with more complicated situations it is often vital to distinguish clearly between those forces acting *on* a particular body and those exerted *by* that body on another. Consider the situation shown in fig. 2:9 where a block rests on a table. The diagram only shows the forces acting

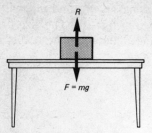

Fig. 2:9 Forces acting on a block at rest on a table.

on the block. The block experiences a gravitational force F and because it has weight it exerts a force *on* the table (the action). The table will exert an equal opposite force *on* the block, which is shown as R and is called the **normal reaction**. (The term "normal" here means that this force is *perpendicular* to the line of contact between block and table.) The fact that we can write

$$mg = R$$

for this situation is simply because the block is in equilibrium, with no net force acting on it. It is not a consequence of the third law since the two forces that we have equated act on the same body. Action and reaction are equal and opposite whether there is equilibrium or not.

Example 2:3

A car of mass 900 kg is travelling along a level road at 15 m/s when the driver removes his foot from the accelerator pedal. If the car comes to rest in a distance of 800 m calculate the average resistive (frictional) force acting on the car.

Solution To find the force we need to determine first the acceleration (which will be negative in this case).

Using equation (2:3) $\qquad v^2 = u^2 + 2as$

$\qquad 0^2 = 15^2 + 2 \times 800a$

$\therefore \quad 1600a = -225 \quad$ giving $\quad a = -\dfrac{225}{1600} = -0.141 \text{ m/s}^2$

We can now find the force using equation (2:5):

$$F = ma = -900 \times 0.141 = -127 \text{ N}$$

The minus sign indicates that the force acts to retard the motion.

2:5 Vectors

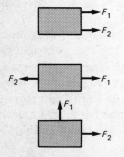

Fig. 2:10 Effect of forces depends on the direction in which they act. The motion in each of the three cases will be different.

Before we proceed to look at circular motion it is worthwhile to bring out a distinction between two types of physical quantity. The first type consists of those quantities which just have **magnitude**. For example to talk of a car of mass 900 kg or a time interval of 20 s fully specifies the quantity. These quantities are called **scalars** and only have a magnitude (which, of course, includes the appropriate unit).

The second type of physical quantity is only fully specified when a **direction** as well as a magnitude is indicated. For example we have talked about the direction of the frictional force and of the acceleration due to gravity. These quantities are called **vectors**. The direction in which they act is a vital piece of information. Fig. 2:10 shows the same two forces acting on the same block but in different directions. The resulting motion will be different in each case.

We could ask whether momentum and energy are vector or scalar quantities. Newton's second law relates momentum and force, and talks about a change of momentum in the direction of the force, which suggests that momentum, like force, is a vector. Energy however is a scalar. We

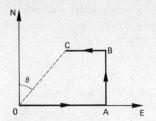

Fig. 2:11 Distinction between distance travelled (solid line) and displacement (dashed line).

will consider work and energy in more detail in chapter 4 and see then that the energy possessed by a moving body depends solely on how fast it is moving, not the direction in which it is moving.

A more rigorous treatment of motion would require us to make a distinction between distance and displacement and between speed and velocity. The term "distance" is used to denote how far a body has travelled from some origin, while "displacement" refers to distance *in a particular direction*. A body which travels along the path OABC in fig. 2:11 has covered a total distance

$$OA + AB + BC$$

but its total displacement is OC at an angle θ to the north. In discussing the equations of motion above we considered distances moved along a particular straight line and so used the term displacement.

In a similar way the term "speed" is used to denote the rate of movement, while "velocity" refers to rate of movement in a particular direction. The body whose motion is shown in fig. 2:11 may move along the path OABC with constant speed but its velocity along the section AB is different from that along OA since the direction has changed. Hence distance and speed are scalar quantities, while displacement and velocity are vectors. Table 2:1 lists some important vectors and scalars.

This distinction between vectors and scalars is important when quantities are to be added together. Scalars are added algebraically so that, for example, masses of 2 kg, 6 kg and 8 kg combine to give a total mass of 16 kg. Similarly an object being lifted by a motor may have a potential energy of 100 J and a kinetic energy of 20 J and hence have a total energy of 120 J.

When vectors are added, account has to be taken of the direction as fig. 2:10 has suggested. In simple cases, addition (or composition as it is sometimes called) can be done mathematically, while in more complex cases a scale drawing may be more appropriate. We will consider a simple case by way of an example.

Table 2:1 Scalar and vector quantities

SCALAR	VECTOR
Distance	Displacement
Speed	Velocity
Mass	Momentum
Energy	Force
Time	Acceleration

Fig. 2:12 Forces on the block in example 2:4.

Example 2:4

A block experiences two forces, of 20 N and 10 N, acting at right angles as shown in fig. 2:12. What is the resultant force?

Solution A vector may be represented diagrammatically by a line drawn in the

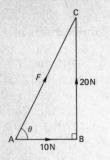

Fig. 2:13 Vector addition of forces shown in fig. 2:12.

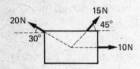

Fig. 2:14 Three coplanar forces acting on a body.

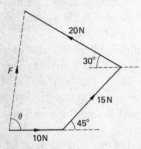

Fig. 2:15 Addition of forces shown in fig. 2:14 by scale drawing.

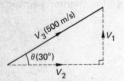

Fig. 2:16 Vector V_3 composed of a horizontal component V_2 and a vertical component V_1.

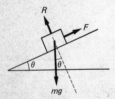

Fig. 2:17 Forces acting on a block on a rough inclined plane.

direction of the vector, with length proportional to the magnitude. Addition is achieved by joining the vectors sequentially as shown in fig. 2:13. The resultant, that is the single force equivalent to the two original forces, has a magnitude AC and acts at an angle θ to the 10 N force.

Since this is a right-angled triangle we can write

$$AC^2 = AB^2 + BC^2$$

i.e. $F^2 = 10^2 + 20^2 = 100 + 400$

giving $F = 22.4 \text{ N}$

From the triangle we also have

$$\tan \theta = \frac{20}{10} \quad \text{giving} \quad \theta = 63.4°$$

The two forces shown in fig. 2:12 are therefore equivalent to a single force of 22.4 N at 63.4° to the direction of the 10 N force.

A similar procedure may be used to add any other vector quantities. Where there are more than two vectors involved a scale drawing is convenient. Fig. 2:14 shows a situation where three forces act on a body. Fig. 2:15 illustrates the use of a scale drawing to determine the resultant. The lines representing the vectors are drawn in turn to give a resultant shown as a dashed line, which corresponds to a force of 20.9 N at an angle θ of 81°. (It has been assumed that the forces all lie in the same plane and that their lines of action pass through the centre of the body so that there is no rotation.)

Addition of vectors reduces them to a single equivalent called the resultant. The reverse procedure, called **resolution of a vector**, enables us to replace a single vector by two or more equivalent vectors. It is often convenient to resolve into two vectors at right angles. For example, a projectile travelling at 500 m/s at 30° to the horizontal could have its motion resolved into two velocities, one horizontal and the other vertical as indicated in fig. 2:16. Velocities V_1 and V_2 when added give V_3 as the resultant. V_1 and V_2 are called the **components** of V_3 and it follows from the geometry of the situation that

$$V_2 = V_3 \cos \theta \qquad V_1 = V_3 \sin \theta$$

and also $V_3^2 = V_1^2 + V_2^2$

For the projectile the horizontal component of velocity is

$$500 \cos 30° = 433 \text{ m/s}$$

and the vertical component is

$$500 \sin 30° = 250 \text{ m/s}$$

As a further illustration of the use of components, consider the block resting on a rough inclined plane shown in fig. 2:17. The frictional force F prevents the block from slipping down the plane. Since there is no motion in any direction the three forces must have zero resultant and, furthermore, the forces in any one direction must balance. It is convenient here to resolve the weight mg into two components (one along the plane and

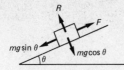

Fig. 2:18 The same situation as in fig. 2:17 with the weight resolved into components along and normal to the plane.

the other normal to the plane) as shown in fig. 2:18. The equilibrium condition can then be written in terms of the forces along the plane and normal to it. Equating forces along the plane gives

$$mg \sin \theta = F$$

while equating forces normal to the plane gives

$$mg \cos \theta = R$$

2:6 Angular Velocity

The motion of a satellite orbiting the earth which was mentioned earlier in section 2:2 is one example of motion along a curved path. We will here consider cases where the path is circular. Common examples of such **circular motion** include the rotation of a gramophone disc, the turning of a wheel, and the movement of a variety of fairground amusements. On the Bohr model of the atom the electron is pictured as orbiting the nucleus in a circular path (see section 19:3). One obvious feature of circular motion is its **periodicity**: the motion repeats itself, usually at a regular rate. If the body rotates at a steady rate, we can define a quantity called the *period* which is the time taken for one complete rotation. The motion then repeats itself each period.

Consider the rotation shown in fig. 2:19 where a body initially at A moves along the arc of a circle to B in a time t and sweeps out an angle θ. Rather than measure the angle in degrees it is convenient to use RADIAN measure. The angle θ in radians is defined as the ratio of arc length to radius, i.e.

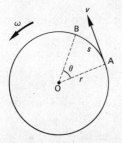

Fig. 2:19 Motion in a circle.

$$\theta = \frac{s}{r} \quad \text{so that} \quad s = r\theta$$

For one complete revolution the arc length s becomes the circumference and consequently

$$\theta = \frac{2\pi r}{r} = 2\pi$$

and we therefore have the identity

$$2\pi \text{ radians} = 360 \text{ degrees}$$

and 1 radian is 57.296°.

Another useful feature of radian measure is that, when small angles are involved (less than 5°), some useful approximations may be used. When the angle θ is measured in radians it is found that

$$\sin \theta \approx \tan \theta \approx \theta \quad \text{and} \quad \cos \theta \approx 1$$

(You could use a calculator or mathematical tables to check the reliability of these small angle approximations.)

To return now to fig. 2:19, when the body is at A its instantaneous direction of motion is along the tangent to the circle, which defines the direction of its velocity. If the body rotates at a uniform rate the speed will not change and we can write

$$s = vt$$

but since from above $s = r\theta$, then

$$r\theta = vt \quad \text{or} \quad \frac{\theta}{t} = \frac{v}{r}$$

The rate of rotation may be specified by defining an **angular velocity** ω as the angle turned through per unit time, i.e.

$$\omega = \frac{\theta}{t} = \frac{v}{r}$$

or $\quad v = r\omega$ (2:6)

With angles in radians, the unit of ω will be rad/s.

Since the rate of rotation may sometimes be changing, the angular velocity is more generally defined as the rate of change of angle and written, using calculus notation, as

$$\omega = \frac{d\theta}{dt}$$

For uniform rotation with angular velocity ω, the time to turn through 2π radians (one rotation) gives the **period**:

$$T = \frac{2\pi}{\omega}$$

Since this is the time for one revolution, the reciprocal of T gives the number of revolutions per second, or the **frequency** f, so that

$$f = \frac{1}{T}$$

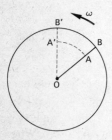

Fig. 2:20 shows a rotating disc. In a short interval of time the line OAB rotates to OA'B' and it follows that all points in the body have the same angular velocity. The speed v however will vary with position, and be a maximum at the edge of the disc (point B).

In many examples of rotational motion the angular velocity will not be constant. For example, if a car accelerates away from a set of traffic lights, the angular velocity of the wheels increases. The wheels therefore undergo an angular acceleration. By analogy with linear acceleration a, we can define an **angular acceleration** α as the rate of change of angular velocity ω so that

$$\alpha = \frac{d\omega}{dt}$$

Fig. 2:20 Rotation of a disc showing the movement of two points A and B on the disc.

which, using $v = r\omega$, becomes $\quad \alpha = \frac{1}{r}\frac{dv}{dt}$

and hence $\alpha = \dfrac{a}{r}$ or $a = r\alpha$

In rotational motion it is important to distinguish between angular and linear velocities and accelerations.

Example 2:5

What is the speed of a point on the edge of a long-playing record rotating at 33 rpm?

Solution Long-playing records have a diameter of 30 cm so the radius is 0.15 m. 33 rpm (revs per minute) corresponds to an angular velocity of

$$\omega = \frac{33 \times 2\pi}{60} = 3.456 \text{ rad/s}$$

The speed at the edge is given from equation (2:6)

$$v = r\omega = 0.15 \times 3.456$$

i.e. $v = 0.52$ m/s

2:7 Centripetal Force

It was indicated earlier in section 2:2 that, to maintain a circular path, an inward force is necessary. In the case of a satellite orbiting the earth the force is provided by the gravitational attraction of the earth, as discussed in section 2:3. In the Bohr model of the atom the electron is held to the nucleus by the Coulomb (electrostatic) attraction (see sections 13:1 and 19:3). In both cases the attractive force is directed towards the centre of the motion and for this reason is referred to as the **centripetal force**. A similar circular motion will result if a mass attached to the end of a light thread is whirled round in a horizontal circle. In this case the tension provides the force towards the centre.

Consider the situation shown in fig. 2:21 where a body at A moves with uniform angular velocity ω through a small angle $\Delta\theta$ from A to B along a short arc length in time Δt. (The symbol Δ here is used to denote a small change in a quantity such as angle, velocity or time.) In moving from A to B, the velocity changes direction but not magnitude. To produce this change in direction a small velocity Δv must be added to the original velocity as shown in fig. 2:22. Since, in this figure, the angle $\Delta\theta$ is small, Δv is approximately the same as an arc length and so we can write

$$\Delta\theta \approx \frac{\Delta v}{v} \quad \text{or} \quad \Delta v \approx v\,\Delta\theta$$

which gives the change in velocity in time Δt. The rate of change is therefore

$$\frac{\Delta v}{\Delta t} \approx \frac{v\,\Delta\theta}{\Delta t}$$

In the limiting situation where $\Delta\theta$ becomes very small, the direction of Δv

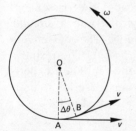

Fig. 2:21 Change in the direction of velocity in moving along a short arc AB.

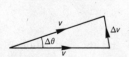

Fig. 2:22 Vector addition showing that ΔV is the change in velocity in moving along the arc AB in fig. 2:21.

in fig. 2:22 becomes normal to v and, since the latter is directed along the tangent, the change in velocity takes place towards the centre. There is therefore a centripetal acceleration given by the limiting value of $\Delta v/\Delta t$ or $v\,\Delta\theta/\Delta t$. The expression for the **centripetal acceleration** is therefore

$$v\frac{\mathrm{d}\theta}{\mathrm{d}t} = v\omega$$

By substituting from equation (2:6) above, this centripetal acceleration can also be written as

$$\frac{v^2}{r} = r\omega^2$$

The centripetal force is found by multiplying this acceleration by the mass m of the body, hence

$$F = \frac{mv^2}{r} = mr\omega^2$$

You should note that the term "centripetal" refers to the direction of the force rather than its nature. The nature of the force depends on the physical situation and table 2:2 summarises four particular cases. The centripetal force produces an acceleration which does not affect the speed of the body but produces a change in the direction of the velocity. Similarly the centripetal acceleration produces no change of angular velocity ω and is quite separate from any angular acceleration α which may additionally be experienced by the body.

A somewhat different effect associated with circular motion is experienced by passengers in a car which turns sharply. Passengers are thrown outwards across the car. This effect arises not because the passengers have changed direction but because the car has turned into a circular path. Fig. 2:23 shows the effect of the car turning. The passengers continue to move in a straight line while the car turns left. The result is that the passengers move outwards relative only to the car. Their experience of an outward motion arises because they are describing their motion from *within* the car which is moving along a curved path.

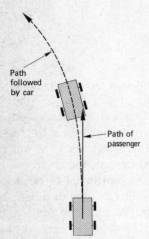

Path followed by car

Path of passenger

Fig. 2:23 Showing how a car passenger is thrown across the car as it turns a corner.

Example 2:6

A satellite rotates in a circular orbit 250 km above the earth's surface where the acceleration due to gravity is 9.09 m/s². Find the period of rotation given that the radius of the earth is 6400 km.

Solution The satellite is above the surface of the earth and the gravitational attraction will therefore be less than that at the surface. Consequently the acceleration due to gravity g also has a lower value than that at the surface. It is this acceleration towards the centre of the earth which provides the centripetal acceleration of the satellite. So we can equate

$$r\omega^2 = g$$

Table 2:2 Examples of circular motion

TYPE OF MOTION	NATURE OF CENTRIPETAL FORCE	EQUATION FOR FORCE
Satellite in orbit	Gravitational attraction	$F = \dfrac{GMm}{r^2} = mg$
Mass on the end of a string	Tension in the string	$F = T$
Electron in orbit round a nucleus	Electrostatic attraction (see section 13:1)	$F = \dfrac{Qq}{4\pi\epsilon_0 r^2}$
Charge moving normal to a uniform magnetic field	Magnetic force (see section 15:6)	$F = BQv$

where r is the distance from the centre of the earth, i.e. $(6400+250)$ km or 6.65×10^6 m. Hence

$$\omega^2 = \frac{g}{r} = \frac{9.09}{6.65 \times 10^6} \qquad \therefore \omega = 1.169 \times 10^{-3} \text{ rad/s}$$

The period of rotation is given by

$$T = \frac{2\pi}{\omega} = \frac{2\pi}{1.169 \times 10^{-3}} = 5374 \text{ s} \quad \text{or} \quad 89.6 \text{ min}$$

Exercises

Where necessary take the acceleration due to gravity $g = 9.8$ m/s^2.

2.1 A car accelerates uniformly from rest for 12 seconds and reaches a final velocity of 15 m/s. Determine the acceleration of the car and the distance travelled. The car then continues to move at constant velocity. Calculate the total time to reach a point 270 m from its starting point.

2.2 A train which is accelerating uniformly at 1 m/s^2 passes two bridges 250 m apart. If its velocity as it passes the first bridge is 20 m/s, calculate the velocity of the train when it reaches the second bridge and the time it takes the train to travel between the bridges.

2.3 A car which is being driven in a town centre accelerates away from a set of traffic lights, reaches a steady velocity of 20 m/s before slowing down again at the next set of lights. Sketch the form of the velocity/time graph for the motion.

2.4 A stone is thrown upwards with a velocity of 25 m/s at an angle of 60° to the horizontal. Determine the horizontal and vertical components of its velocity.
Explain how gravity will influence the subsequent motion of the stone. (It is helpful to consider the two components of the motion separately.)

2.5 A block of mass 4 kg slides down a rough plane inclined at 30° to the horizontal at constant velocity. Calculate the frictional force between the block and the plane.

2.6 A body of mass 20 kg experiences two forces of magnitudes 12 N and 15 N acting at right angles. Calculate the acceleration of the body. Note that acceleration is a vector so you also need to indicate the direction of motion. Determine the angle between the direction of motion and the 15 N force.

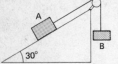

2.7 A block A of mass 1 kg on a rough plane inclined at 30° to the horizontal is attached to a light string which passes over a light frictionless pulley at the top of the incline. The other end of the string carries a mass B of 2 kg as shown in fig. 2:24. Draw a diagram showing all the forces acting on block A.

Fig. 2:24

2.8 When the blocks described in exercise 2.7 are released it is found that block A accelerates up the plane at 3 m/s². Calculate the tension in the string and the frictional force between block A and the plane.

2.9 The forge head of a forging machine has a mass of 1500 kg. The head falls onto a metal block and strikes it with an impact velocity of 2 m/s. The head is then brought to rest in a distance of 50 mm. Calculate the mean retarding force on the head as it decelerates.

2.10 A car is travelling at 25 m/s. If the wheels are of radius 0.3 m how many revolutions do the wheels make per second?

2.11 If the car in exercise 2.10 reached its velocity from rest in 15 s, determine the average angular acceleration of the wheels.

2.12 Calculate the angular velocity of a car which is driven round a corner of radius 90 m at a speed of 15 m/s. Determine the centripetal acceleration of the car during the turn.

2.13 A geostationary satellite in a circular orbit round the equator of the earth rotates at the same angular velocity as the earth and so remains stationary with respect to the ground. If its orbital radius is 42 300 km, calculate the centripetal acceleration of the satellite. (This will also be the gravitational acceleration at this distance from the earth.)

2.14 An electron of mass 9.1×10^{-31} kg moves in a circular path at right angles to a magnetic field under a centripetal force of 1.2×10^{-15} N. Given that the speed of the electron is 6×10^{6} m/s, determine the radius of the circular path.

3 Moments and Equilibrium

In chapter 2 we found that a linear accelerating motion resulted from the action of a resultant force on a body. We will now consider situations where forces act to produce a rotational acceleration. It was also seen that, where the forces acting on a body balanced, it would move at constant velocity or remain at rest. In this latter case the forces act to maintain an equilibrium, that is the body does not roll, topple or move in any other way. These equilibrium conditions are important in, for example, the design of mechanical structures and will be examined in this chapter.

3:1 Moment of a Force

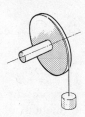

Fig. 3:1 Angular acceleration of a wheel and axle due to a mass attached to a thread wound round the rim. The dotted line is the axis of rotation.

In the discussion of linear motion it was assumed that the shape of the body in question did not significantly affect the motion. An extended body, such as a car or ball, was treated as if all its mass was concentrated at a point, which is called the centre of mass. It was also assumed that the forces acted through this centre of mass.

These conditions are not always met. In rotational motion the angular acceleration of a body under a given force is found to depend not only on the mass but also on the shape of the body. Furthermore the forces do not generally act through the centre of mass.

Fig. 3:1 shows a wheel and axle which are free to turn. The arrangement can be accelerated by a mass attached to a thread which is wound round the rim of the wheel. If the mass is released, it falls, the thread unwinds, and the wheel begins to turn. The mass accelerates linearly downwards and the wheel undergoes an angular acceleration.

The **turning effect** of the force shown in fig. 3:2 is measured by the moment of the force, T, which is defined by the product

$$T = Fr$$

The unit for the moment of a force is therefore newton metres (N m). The moment of a force is frequently called the **torque** which is the term we shall generally use.

In a more general case such as that shown in fig. 3:3, where an irregular body is suspended about an axis through O, the torque is given by the product of the force and the perpendicular distance from the point of suspension to the line of action of the force, shown as r in the figure. In both figs. 3:2 and 3:3 the rotation will occur in a clockwise direction and the torque is said to have a *clockwise sense*. If the direction of F was reversed, the torque and the resulting rotation would be anticlockwise.

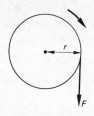

Fig. 3:2 Force at distance r from axis produces rotation in sense indicated by arrow.

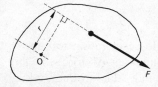

Fig. 3:3 Torque of force on an irregular body. Force acts along a line at perpendicular distance r from suspension point O.

3:2 Rotational Motion

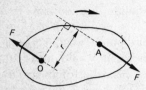

Fig. 3:4 Reaction force at suspension produces a couple. The arrow indicates the sense of rotation.

When a body suspended at a fixed point, such as in fig. 3:3, is acted on by a force, pure rotation results. However to achieve this, a second force, equal and opposite to the force F, is needed to act on the body as shown in fig. 3:4. Otherwise linear motion would also occur. This second force is supplied by the reaction on the body at the point of suspension, and consequently acts through O. The system of forces in fig. 3:4, with equal and opposite forces acting along different, but parallel, lines of action, is called a **couple**. It is a couple which produces pure rotation. The torque T of the couple about O is

$$T = Fr$$

since the force through O has no turning effect about O.

When a torque acts on a body it produces rotational motion with an angular acceleration α (which was met in section 2:6). The relationship between the torque and the angular acceleration can be expressed in an equation similar to the linear equation between force and acceleration:

$$F = ma$$

In the rotational case, the angular acceleration α is proportional to the applied torque, i.e. with I as the constant

$$T = I\alpha$$

where I is called the **moment of inertia** of the body. Just as m represents the inertia or opposition of a body to linear motion, so the moment of inertia I represents the opposition to rotational motion.

The value of the moment of inertia for a given body depends not only on the mass m of the body but also on the size or shape of the body; in other words on the way in which the mass is distributed.

Example 3:1

A disc of moment of inertia 10^{-2} kg m^2 is suspended so that it is free to turn about an axis through its centre perpendicular to its plane. The disc is accelerated by a force of 0.5 N acting tangentially at its rim. Calculate the angular acceleration of the disc if the radius of the disc is 0.1 m.

Solution The force acting at the rim exerts a torque, as indicated in fig. 3:2, of magnitude

$$T = Fr = 0.5 \times 0.1 = 0.05 \quad \text{or} \quad 5 \times 10^{-2} \, \text{N m}$$

To find the angular acceleration we can substitute into the equation of motion:

$$T = I\alpha \quad \text{giving} \quad 5 \times 10^{-2} = 1 \times 10^{-2}\alpha$$

so that $\quad \alpha = \dfrac{5 \times 10^{-2}}{1 \times 10^{-2}} = 5 \, \text{rad/s}^2$

3:3 Equilibrium

In example 3:1, the disc rotated under a single accelerating torque. In other situations a body may experience a number of torques. The resultant torque is then found by combining the individual torques, taking

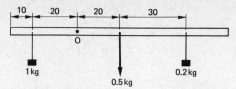

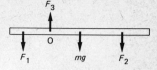

Fig. 3:6 Forces acting on the bar in fig. 3:5.

Fig. 3:5 Massive bar pivoted at O carrying two loads. (Distances are in centimetres.)

account of whether they act in a clockwise or anticlockwise sense. It is also possible that the resultant may be zero so that no motion, or change of motion, occurs.

Fig. 3:5 shows a bar of length 1 m and mass 0.5 kg which is pivoted 30 cm from one end. The weight of the bar acts at its centre as shown and so produces a clockwise torque about O of

$$0.5 \times 9.8 \times 0.2 = 0.98 \text{ N m}$$

The bar also carries loads of 1 kg and 0.2 kg positioned as shown. We want to determine the resultant torque on the bar.

The 0.2 kg load produces a clockwise torque of

$$0.2 \times 9.8 \times 0.5 = 0.98 \text{ N m}$$

which added to the torque of the weight gives a total clockwise torque of

$$0.98 + 0.98 = 1.96 \text{ N m}$$

The 1 kg load will exert an anticlockwise torque of magnitude

$$1 \times 9.8 \times 0.2 = 1.96 \text{ N m}$$

Since this anticlockwise torque just balances the total clockwise torque, the net torque is zero and the bar will remain **in equilibrium**.

We have here an example of the law which is often called the **principle of moments**:

When a body is in equilibrium, the sum of the clockwise moments (or torques) is equal to the sum of the anticlockwise moments (or torques).

If this principle is satisfied, then there will be no rotation.

The condition for no linear motion has been met earlier in section 2:5, where we looked at a block on a rough plane. Figs. 2:17 and 2:18 show the force diagrams. The three forces balance and the block remains at rest. Furthermore the components of the forces in any direction also give zero resultant. Since the loaded bar in fig. 3:5 is also in equilibrium, the forces acting on it must balance. Fig. 3:6 shows all the forces acting on the bar. The pivot exerts a reaction force F_3 which balances the downward forces due to gravity. We can then write

$$F_3 = F_1 + F_2 + mg$$

Notice also that F_3 will have no turning effect about O.

This example of a bar in equilibrium under a system of forces illustrates the general conditions which must be met for any body to be in equilibrium:

1　The net force along any axis is zero.
2　The net torque about any axis is zero.

If these two conditions are met, the body will undergo neither linear nor angular acceleration.

Example 3:2

Fig. 3:7(a) shows a uniform bar AB 1.5 m long of mass 6 kg resting on two supports at C and D. Calculate the two reaction forces at the supports.

Fig. 3:7 (a) Bar resting on two supports (distances in m). (b) Forces on the bar.

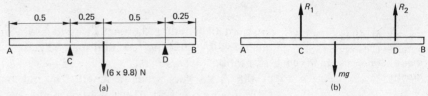

(a) (b)

Solution The bar is in equilibrium under three forces: its weight and the two reaction forces R_1 and R_2 at the supports, all of which are shown in fig. 3:7(b). We can apply the two conditions for equilibrium in turn. Since there is no linear motion the net force vertically is zero and so

$$R_1 + R_2 = mg = 6 \times 9.8 = 58.8 \, \text{N} \tag{1}$$

The bar does not rotate so we can apply the second condition for equilibrium which is the principle of moments. If we take torques about the end A, then R_1 and R_2 produce anticlockwise torques while the weight produces a clockwise torque. For equilibrium

$$(R_1 \times 0.5) + (R_2 \times 1.25) = 6 \times 9.8 \times 0.75 = 44.1$$

and multiplying each term by 2 gives

$$R_1 + 2.5R_2 = 88.2 \tag{2}$$

R_1 and R_2 can be found from these simultaneous equations (1) and (2). Subtracting the two equations gives

$$2.5R_2 - R_2 = 88.2 - 58.8$$
$$1.5R_2 = 29.4$$

so that $\quad R_2 = 19.6 \, \text{N}$

Substituting this value back into equation (1) gives

$$R_1 + 19.6 = 58.8 \quad \text{so} \quad R_1 = 58.8 - 19.6 = 39.2 \, \text{N}$$

Example 3:3

A ladder of length 8 m and mass 25 kg has its centre of gravity 3 m from its base. It rests with one end on rough ground and the other end against a smooth vertical wall, with the ladder inclined at 60° to the horizontal. Calculate the reaction forces and the frictional force at the ground.

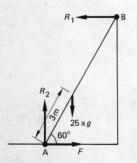

Fig. 3:8 Forces on a ladder resting against a smooth wall.

Solution Fig. 3:8 shows the forces acting on the system. The only frictional force is that between the ladder and the ground (force F) since the wall is smooth. There will be two reaction forces, R_2 normal to the ground and R_1 normal to the wall. The only other force is the weight of the ladder.

The two general equilibrium conditions can be applied in turn.

Equating the horizontal forces gives $\quad R_1 = F$

Equating vertical forces: $\quad R_2 = 25 \times 9.8 = 245 \, \text{N}$

Torques may be taken about any point in the plane. If point A is used then the reaction force at the wall has an anticlockwise torque about A of

$$R_1 \times 8 \cos 30°$$

since the ladder is 8 m long.

The weight produces a clockwise torque of

$$25 \times 9.8 \times 3 \cos 60° = 367.5 \text{ N m}$$

so that, in equilibrium, equating clockwise and anticlockwise torques gives

$$R_1 \times 8 \cos 30° = 367.5$$

$$R_1 = \frac{367.5}{8 \times 0.866} = 53.04 \text{ N}$$

The frictional force F is therefore also equal to 53.04 N.

3:4 Centre of Gravity

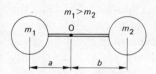

Fig. 3:9 Balls joined by a light rod (O is the centre of gravity).

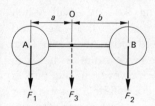

Fig. 3:10 Gravitational forces on the balls in fig. 3:9. F_3 is the equivalent force at the centre of gravity.

When we dealt earlier with linear motion we treated an extended body as if all the mass were concentrated at one point. The position of this equivalent point mass is called the **centre of gravity**†. As a simple example, consider the two balls, of masses m_1 and m_2, joined by a light (massless) rod as shown in fig. 3:9.

The position of the centre of gravity may conveniently be found by considering the effect of gravitational forces on the system. Each ball in fig. 3:9 will experience a force as indicated in fig. 3:10. The centre of gravity is the position at which a single force F_3, equivalent to the combined effect of F_1 and F_2, would act. For F_3 to be the equivalent force its magnitude must be given by

$$F_3 = F_1 + F_2$$

while the position at which F_3 acts can be determined using the principle of moments.

If we take torques about A (the centre of mass m_1) then
the torque of F_2 is $\quad F_2(a+b)$ (clockwise)
and the equivalent torque due to F_3 is $\quad F_3 a$ (clockwise)
For equivalence $\quad F_3 a = F_2(a+b)$
Substituting $F_3 = F_1 + F_2$ gives

$$(F_1 + F_2)a = F_2(a+b)$$

$$F_1 a + F_2 a = F_2 a + F_2 b$$

or $\quad F_1 a = F_2 b$

Since $F_1 = m_1 g$ and $F_2 = m_2 g$ this reduces to

$$m_1 a = m_2 b$$

In simple cases the location of the centre of gravity is fairly obvious. For example, it occurs half way along a uniform rod and at the geometrical centre of a sphere. In the case of a rectangular plate the centre of gravity will be at the intersection of the diagonals.

For irregular shapes the centre of gravity may be located experimentally. The method for an irregular flat plate such as that shown in fig. 3:11 involves supporting it about a horizontal axis at two different points in

† If all the mass of a body is concentrated at a point, called the centre of mass (p. 19), the gravitational force will act through this point. Hence it is commonly called the centre of gravity.

Fig. 3:11 Locating the centre of gravity G for an irregular body.

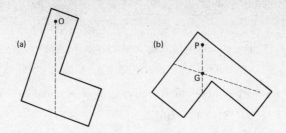

turn. The method assumes that the plate is of uniform, but small, thickness. Such a body is called a **lamina**. The laminar plate is first suspended from O as in fig. 3:11(a), and a vertical line drawn through O using a plumbline. The centre of gravity will lie somewhere along this line. The plate is then suspended about a second point P and a second line drawn. The centre of gravity must also lie along this line and the only possible position is therefore point G.

3:5 The Stability of Equilibrium

We have found two general conditions for the equilibrium of a body: the absence of any net force and the absence of any net torque. We can now go further and consider the **stability of the equilibrium** in different cases. To do this we consider what happens to a body when it is slightly displaced from its equilibrium position. A pencil standing on its end for example will immediately fall over if it is displaced from the vertical, while a child's "kelly" toy is designed so that if it is pushed to one side it returns to its original position. This latter situation is an example of stable equilibrium while the pencil was originally in unstable equilibrium.

These two situations are further illustrated in fig. 3:12 by a ball. When the ball is in equilibrium, the forces on it balance so that the weight and normal reaction force will be oppositely directed along the same line. When the ball is displaced in case (c), the same force situation holds in the new position and the ball remains there. However if the ball is displaced from unstable equilibrium (case (b)), there will be a net force on it which will move the ball further away as fig. 3:13(a) shows. On the concave surface however (case (a)), the displacement leads to a restoring force as shown in fig. 3:13(b), and stable equilibrium results.

Finally we look at the toppling of bodies. Fig. 3:14 shows the forces on a block on a rough plane at two different inclinations. In each case friction prevents sliding. In case (a) the weight and the frictional force act through point A, and since for equilibrium there can be no torque about A the reaction force must also pass through A. (This illustrates a general equilibrium condition under three forces in the same plane: that the forces must have lines of action passing through a common point, i.e. the forces are **concurrent**.) In case (b) with greater inclination, the frictional force and weight again pass through A but, since it is now outside the body, the reaction force cannot pass through A and there will be a net torque which will cause the block to tip.

In a similar way a bus on an inclined surface (as in fig. 3:15) will not tip

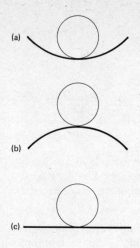

Fig. 3:12 Types of equilibrium: (a) stable, (b) unstable, (c) neutral.

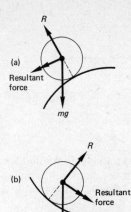

Fig. 3:13 Forces acting on
the ball in fig. 3:12 when
displaced: (a) unstable,
equilibrium, (b) stable
equilibrium.

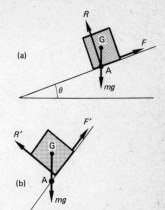

Fig. 3:14 Forces on a block:
(a) in equilibrium, (b) condition for tipping (point A lies
outside area of contact).

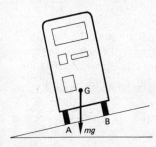

as long as the line of action of the weight passes between the wheels A and B. To meet this condition the vehicle is designed with its centre of gravity as low as possible.

Fig. 3:15 Bus on a inclined
plane. For stability, weight
must act between wheel
base.

Exercises

Where necessary take the acceleration due to gravity $g = 9.8 \text{ m/s}^2$

3.1 A jib crane has an arm 15 m long inclined at 60° to the horizontal. If the crane lifts a load of 3000 kg (3 tonnes) calculate the turning moment which will tend to topple the crane.

3.2 A beam XY of length 4 m and mass 60 kg is pivotted at end X and held horizontal by a chain attached at end Y. By considering the torques acting at equilibrium calculate the tension in the chain if the chain is inclined at 45° to the horizontal.

3.3 A torque of 200 N m applied to the shaft of a motor accelerates it from rest to a rotational velocity of 300 rev/min after 15 s. Determine the angular acceleration and the moment of inertia of the rotating system.

3.4 A rotating flywheel of moment of inertia 10 kg m^2 and radius 0.5 m is slowed down by a frictional force of 25 N applied at the rim. Calculate the time it takes to come to rest if its initial angular velocity is 250 rad/s.

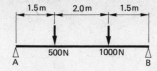

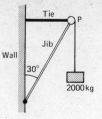

Fig. 3:16 Loaded bar.   **Fig. 3:17** Jib crane.

3.5 A uniform plank of mass 30 kg and length 9 m stands on the side of a swimming pool with 3 m of the plank projecting over the edge. Determine how far a person of mass 60 kg can walk along the plank beyond the edge of the pool before the plank tips.

3.6 A horizontal beam of length 5 m is supported at its two ends. Loads of 500 N and 1000 N act as shown in fig. 3:16. Determine the magnitude of the reaction forces at the supports. The mass of the beam may be assumed to be negligible.

3.7 A uniform rod of mass 1 kg and length 1 m has a small ball attached to each end. If the balls have masses of 250 g and 500 g respectively calculate the position of the centre of gravity from the end carrying the 250 g ball.

3.8 A mass of 2 kg is hung from the ceiling by two light threads of length 1.6 m and 1.2 m respectively. If the angle between the two threads is 90° calculate the tension in each of the supporting threads.

3.9 Fig. 3:17 shows a crane which carries a load of 2000 kg (2 tonnes). By considering the equilibrium of the forces acting at the point P determine the tension in the tie and the force in the jib. State whether the jib is under tension or compression.

3.10 A block of mass 400 g rests on a rough inclined plane. If the frictional force between the block and the plane has a maximum value of 2 N, calculate the maximum angle of inclination to the horizontal for which the block will not slip down the plane.

4 Momentum and Energy

This chapter deals in more detail with the concepts of momentum, work and energy. We shall consider how these concepts relate to the motions discussed in preceding chapters and see how the conservation principles, which are concerned with the conditions under which the momentum or the energy of a system remains constant, can be applied to the analysis of such motions.

4:1 Momentum and its Conservation

In chapter 2 momentum was defined as the product of mass and velocity

$$p = mv$$

and, since velocity is a vector, momentum is also a vector quantity. On the basis of Newton's second law, momentum will have units of newton-seconds (N s). Also in chapter 2 we found that when a net force acts on a body its velocity, and its momentum, change.

Suppose a force F is exerted on a body A by another body B. Force F is equal to the *rate* of change of momentum. If F is constant over a time t then the *total* change of momentum is given by the product Ft. If the force accelerates body A from velocity u to velocity v, the total change of momentum is

$$mv - mu = Ft \tag{4:1}$$

According to Newton's third law however, body A exerts an equal and *opposite* force on body B so that B undergoes the same change in momentum but in a direction opposite to that of A. The total change of momentum of the two interacting bodies is therefore zero. What we have here is a particular example of the principle of the conservation of momentum which is seen to follow directly from Newton's laws.

The principle of the **conservation of momentum** may be stated more generally in the following form:

In a closed system of interacting bodies, the total momentum of the system remains constant.

The term **closed system** means that there are no external forces acting on the system to affect the interaction, for example by preventing one body from moving. To explain the principle we will consider some illustrations.

Fig. 4:1 shows two trucks on a horizontal track. Initially truck B, of mass m_B, is stationary while truck A, of mass m_A, approaches it at velocity u. The initial momentum of the system is therefore

$$p_1 = m_A u$$

If the trucks couple up on collision and then move with velocity v (fig. 4:1(b)) the momentum after impact is given by

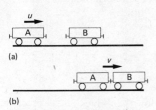

Fig. 4:1 Motion of trucks on a horizontal track: (a) before collision, (b) after collision.

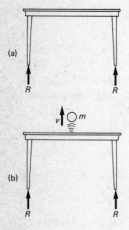

(a)

(b)

Fig. 4:2 Ball bouncing off a table: (a) before collision, (b) after collision. R represents the reaction forces on the table.

$$p_2 = (m_A + m_B)v$$

From the conservation principle the momentum is unchanged by the collision and

$$m_A u = (m_A + m_B)v$$

If the initial conditions are known, the final velocity v can be calculated.

A different situation is shown in fig. 4:2 where a ball of mass m bounces on a horizontal table. If u is the velocity of impact, the momentum immediately before collision is mu, while if v is the rebound velocity the momentum after collision is $-mv$. (The negative sign arises because momentum is a vector and indicates a reversal of direction of motion.) Since the table does not move, the momentum of the system does not remain constant through the collision. The principle cannot be applied here because the system is not closed. The reaction force the floor exerts on the table is an external force on the ball-table system which prevents the latter from moving.

In a collision such as that between two snooker balls, the time of contact is short so that the forces act for only a fraction of a second. Such short-duration forces are called **impulsive forces**. Other examples include a ball hit with a bat and the impact of a hammer or punch. In these cases the product of force and the time for which it acts defines a quantity called the **impulse** I. From equation (4:1) it can be seen that the impulse is also equal to the total change of momentum, since

$$I = Ft \quad \text{and} \quad Ft = mv - mu$$

It is assumed here that the force is constant over the short time t. However even in more general cases where the force varies, the impulse is still given by

$$I = mv - mu$$

Example 4:1

A ball of mass 0.25 kg rolling across a flat table at 4 m/s catches up and collides with a second ball of mass 1 kg moving at 2 m/s. After collision the 1 kg ball moves at 2.5 m/s in the same direction. Calculate the final velocity of the first ball.

Solution Fig. 4:3 shows the situation before and after collision. Both balls have an initial momentum in the same direction and the total initial momentum is the sum

$$(0.25 \times 4) + (1 \times 2) = 3 \text{ N s}$$

From fig. 4:3 (b) the total final momentum is

$$(0.25 \times v) + (1 \times 2.5) = 0.25v + 2.5$$

Equating the total momentum before and after collision then gives

$$0.25v + 2.5 = 3$$

hence $0.25v = 3 - 2.5 = 0.5$

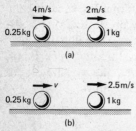

Fig. 4:3 Motion (a) before collision, (b) after collision.

and $v = \dfrac{0.5}{0.25} = 2 \text{ m/s}$

Had the value of velocity v come out negative here, it would have indicated that the ball was moving in a direction opposite to that shown in fig. 4:3(b).

Example 4:2

A radioactive nucleus of mass 226 u is initially stationary. The nucleus splits into an alpha-particle of mass 4 u which travels at 2×10^7 m/s and a residual nucleus of mass 222 u. Describe the motion of this residual nucleus. [The masses here are in unified atomic mass units—see section 20.2.]

Solution The initial momentum of this system is zero and from the principle of the conservation of momentum it follows that the final momentum must also be zero. We can say straight-away that the residual nucleus must move in the opposite direction to the alpha-particle.

The alpha-particle momentum is $4 \times 2 \times 10^7 = 8 \times 10^7$

If the speed of the residual nucleus is v its momentum is $222v$ and, since this is equal and opposite to the α-particle momentum, then

$$222v = 8 \times 10^7 \quad \text{so} \quad v = \frac{8 \times 10^7}{222} = 3.60 \times 10^5 \text{ m/s}$$

The argument used here can be extended to situations where more than two fragments are produced. For example, in an explosion, pieces may fly off in all directions. However the sum of the momentum of the individual pieces will be zero. (Remember that the summing here will involve vector addition.)

4:2 Work, Energy and Power

When a force produces movement, for example in the extension of a wire under an applied load or the launch of a space rocket, **work** is done. The amount of work done depends on the distance moved under the application of the force.

Suppose that a constant force F acts on a body which moves in a direction at an angle θ to F as shown in fig. 4:4. If the body moves distance s the work done is given by

$$W = Fs \cos \theta$$

i.e. work done is distance times component of the force in the direction of motion. In the simpler, and more common, case where the body moves in the direction of the force, θ is zero, and so we have

$$W = Fs \quad \text{or} \quad \text{Work done} = \text{force} \times \text{distance}$$

The unit of work will be a newton metre and is called the JOULE (J), i.e. $1 \text{ N m} = 1 \text{ J}$.

The above equations can only be used if the force is constant. In the more general situation where force changes with distance, use can be made of a force/distance graph. For example, a rocket travelling away from the earth experiences a weakening gravitational force as shown in fig. 4:5. (The force/distance relationship here was discussed in section 2:3.) The work done in travelling from distance r_1 to r_2 is determined by

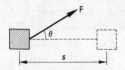

Fig. 4:4 Force F producing motion over distance s does work.

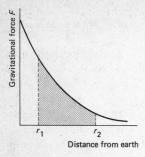

Fig. 4:5 Force/distance graph for a rocket showing how the gravitational force decreases with increasing distance from the earth.

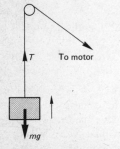

Fig. 4:6 Work is done to raise a load against gravity.

measuring the area under the graph (which is shown shaded). This procedure will be used to determine the work done in stretching a wire in chapter 9 where fig. 9:11 is the appropriate force/distance graph. In general then, work done is given by the area under a force/distance graph.

When work is done it is accompanied by a change in energy. **Energy** is usually thought of in relation to work, and a body or system which possesses energy is capable of doing work. When a load is lifted over a pulley as shown in fig. 4:6, the tension in the cable has to be sufficient to overcome the gravitational force (the weight). Work is done *on* the load that is being raised and so its energy *increases*. On the other hand, a block sliding across a rough surface will eventually be brought to rest because of friction between the block and the surface. Here work is done *by* the block against the frictional force and the block *loses* energy.

These examples illustrate the general rules:
1   When work is done *by* a body, it loses energy.
2   When work is done *on* a body, it gains energy.
When work is being done continuously it is useful to know the rate of working, i.e. the work done per unit time. This quantity is called the **power**. If W is the work done over a time interval t, the average power P_{av} is

$$P_{av} = \frac{W}{t}$$

and its unit, which is joules/second from this equation, is called the WATT (W). From the definition it follows that the more powerful a machine is, the faster it will work. Note that work, energy and power are all scalar quantities.

4:3 Kinetic and Potential Energy

Energy sources such as nuclear, oil and coal have much in common. For example, whichever is used to run an electricity generating station, the energy stored within the fuel is released, work is done to drive the turbines, and this is in turn converted into electrical energy. The energy stored in the fuel (potential energy), the work done in driving the turbines, and the energy of the mechanical movement (kinetic energy) are closely linked.

In all the different sources, energy is present either as potential or as kinetic energy. Within the atom for example, energy is stored in the nucleus because of the tight grouping of the neutrons and protons (see section 20:3). In a gas, the thermal energy is identified with the kinetic energy of the moving molecules (see section 11:4). In circuits, electrical energy is associated with the kinetic energy of moving charges and this is derived from the original potential energy stored in the battery or accumulator.

1   Kinetic energy (KE) is the energy possessed by a body because of its motion. Consider a body of mass m accelerated from rest by a constant force F so that after travelling a distance s its velocity is v. From the equations of motion (section 2:1), we can write (since the initial velocity $u = 0$)

$$v^2 = 2as \quad \text{or} \quad s = \frac{v^2}{2a}$$

while Newton's second law gives $\quad F = ma$

Multiplying these equations gives

$$Fs = \tfrac{1}{2}mv^2$$

The left-hand side of this equation is the work done *by* the force *on* the body and the right-hand side is the energy associated with the motion. It gives the general expression for the kinetic energy of linear motion with speed v:

$$KE = \tfrac{1}{2}mv^2$$

2 Potential energy (PE) is the energy possessed by a body because of its position. A change in position can result in a change of PE. For example, when a wire is stretched, the molecular separation is slightly increased and so the PE of the molecules increases. (This energy stored in the stretched wire is discussed in section 9:7.) The PE of a stationary body may be released when movement occurs. The energy of a body due to its height above the ground is released as it falls and becomes the kinetic energy of the falling body.

A mass at the earth's surface experiences a gravitational force, which we call the weight, given by

$$F = mg$$

We can also describe this situation by saying that the earth has a **gravitational field** associated with it and that a mass located in this field experiences a force. At a large distance from the earth the attractive force is very small and the gravitational field very weak. The strength of the field is defined as the force per unit mass, i.e. by the ratio F/m, so that

$$\text{Gravitational field strength} = \frac{F}{m} = g$$

It can be seen then that the acceleration due to gravity (or acceleration of free fall), measured in m/s^2, is numerically equal to the gravitational field strength, measured in N/kg.

When a body moves in a gravitational field it generally changes its potential energy. For the body raised through height h in fig. 4:7, the lifting force must be sufficient to just overcome the gravitational force on the body, and the work done on the body by the lifting force is mgh.

Since work is done *on* the body this gives the increase in energy so that, taking the ground as the zero position, at height h above the ground,

$$PE = mgh$$

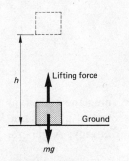

Fig. 4:7 Body lifted through height h gains potential energy mgh.

It is assumed here that h is small so that any change of g over this distance can be ignored.

Work is done in lifting a body against the gravitational force and it is therefore only the vertical component of motion which changes the PE. In a horizontal movement there is no work done against gravity and so no change of potential energy.

Example 4:3

The first-stage rocket of a space probe of mass 2.8×10^6 kg produces a burn lasting 100 sec at the end of which the rocket has reached a speed of 1200 m/s and a height of 60 km. Calculate the changes in kinetic and potential energy during this burn. For the purpose of calculation assume that the mass and the acceleration due to gravity (9.8 m/s^2) remain constant.

Solution   The final velocity is 1200 m/s and so the kinetic energy after 100 sec is

$$\tfrac{1}{2}mv^2 = \tfrac{1}{2} \times 2.8 \times 10^6 \times (1200)^2 = 2.02 \times 10^{12} \text{ J}$$

If the rocket starts from rest this is the gain of kinetic energy in the burn.

The gain in potential energy (taking the earth's surface as the zero) is given by

$$mgh = 2.8 \times 10^6 \times 9.8 \times 60 \times 10^3 = 1.65 \times 10^{12} \text{ J}$$

Note that in practice the mass will decrease as fuel is burnt and also that the acceleration due to gravity decreases as the rocket rises (reaching about 9.62 m/s^2 at 60 km).

Example 4:4

Calculate the average thrust exerted by the rocket during the 100 sec burn in example 4:3.

Solution   We have already calculated the increase in both kinetic and potential energy. Hence over 100 sec the total energy gained by the space probe is

$$\text{KE gained} + \text{PE gained} = (2.02 \times 10^{12}) + (1.65 \times 10^{12}) = 3.67 \times 10^{12} \text{ J}$$

This is the work done by the rocket in lifting the probe 60 km.

Using   work done = force × distance   gives

$$F \times 60 \times 10^3 = 3.67 \times 10^{12}$$

and hence the average thrust is

$$F = \frac{3.67 \times 10^{12}}{60 \times 10^3} = 6.12 \times 10^7 \text{ N}$$

4:4 Conservation of Energy

A second fundamental principle (alongside that concerning the conservation of momentum), which finds application in many areas of physics and other sciences, is the principle of the **conservation of energy**. A general statement of this principle is

> *In a closed system, the total amount of energy remains constant, although the energy may change its form.*

A closed system here means one which is isolated so that no energy

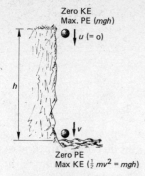

Zero KE
Max. PE (mgh)

$u (= 0)$

h

v

Zero PE
Max KE ($\frac{1}{2}mv^2 = mgh$)

Fig. 4:8 Body falling off a cliff—showing energy conservation.

transfer to it or from it can occur. Within such a system, energy cannot be created or destroyed.

A stone which falls off a vertical cliff moves freely under gravity and in the absence of friction (air resistance) forms a closed system. As the stone falls, it loses potential energy and gains kinetic energy (fig. 4:8). From the principle of energy conservation we conclude that the loss of potential energy will equal the gain of kinetic energy, so, if the initial velocity is zero,

$$mgh = \tfrac{1}{2}mv^2 \quad \text{or} \quad 2gh = v^2$$

In a system where frictional forces act, mechanical energy is dissipated. A body sliding across a horizontal plane against a frictional force will come to rest, so that it loses all its kinetic energy without any change in potential energy. The energy is lost in doing work against friction and the result is that mechanical energy is converted into thermal energy (i.e. heat). The presence of frictional forces always produces heating, which needs to be minimised in a car engine by the use of lubricants but which is utilised in a resistance coil to produce electrical heating.

We have, up to now, thought of energy and mass as quite separate concepts, but modern physics, and in particular the theory of relativity, suggests that this is not always a valid separation (although it works in everyday situations). One of the consequences of relativity theory is that there is an equivalence between mass m and energy E which is summed up in the equation

$$E = mc^2$$

where c is the speed of light (3×10^8 m/s in a vacuum). One example of where mass-energy equivalence is important is in the release of the energy locked up in atomic nuclei.

Example 4:5

A block of mass 4 kg slides down a plane inclined at 30° to the horizontal against a frictional force of 12 N. Determine the kinetic energy at the bottom if the block falls through a vertical height of 0.75 m.

Solution Fig. 4:9 shows the situation and the forces acting on the block. A solution to this problem could be found by looking at the forces but we will use an energy approach.

Since the block moves through a vertical height of 0.75 m the potential energy lost is

$$mgh = 4 \times 9.8 \times 0.75 = 29.4 \text{ J}$$

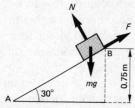

N

F

B

mg

$30°$

A

0.75 m

Fig. 4:9 Forces on a block on an inclined plane.

From fig. 4:9 $\sin 30° = \dfrac{0.75}{AB}$

where AB is the length of the plane and so

$$AB = \frac{0.75}{\sin 30°} = 1.5 \text{ m}$$

The work done against friction in moving 1.5 m against a 12 N force is

$$12 \times 1.5 = 18 \text{ J}$$

From the conservation of energy:

PE lost = KE gained + work done against friction

so that   KE gained is $29.4 - 18 = 11.4 \text{ J}$

4:5 Rotational Energy and Momentum

A rotating body, like a body in linear motion, has kinetic energy since it is in motion. If the body has moment of inertia I and is rotating with angular velocity ω, the expression for the **rotational KE** is

$$KE = \tfrac{1}{2}I\omega^2$$

(You should note the similarity in form between this equation and the linear expression $\tfrac{1}{2}mv^2$.) Two other equations which apply to rotation are also similar to their linear equivalents. When a body rotates through an angle θ (measured in radians), the work done is given by

$$W = T\theta$$

where T is the torque on the body. Finally the **angular momentum** L of a rotating body is given by the expression

$$L = I\omega$$

4:6 Collisions

In section 4:1 we applied the conservation of momentum principle to collisions to determine the resulting motion. But what happens to the kinetic energy in collisions? We can look again at example 4:1 above to calculate the initial and final kinetic energies. The result is

Initial KE $= (\tfrac{1}{2} \times 0.25 \times 4^2) + (\tfrac{1}{2} \times 1 \times 2^2) = 4 \text{ J}$

Final KE $= (\tfrac{1}{2} \times 0.25 \times 2^2) + (\tfrac{1}{2} \times 1 \times 2.5^2) = 3.625 \text{ J}$

In this collision therefore, kinetic energy is lost and such collisions are described as **inelastic**. Cases where kinetic energy (as well as momentum) is conserved are called **elastic** collisions.

It is important to remember that in solving collision problems the conservation of momentum can be applied to all cases, but kinetic energy is only conserved if the collision is elastic. The loss of kinetic energy in inelastic collisions does not, of course, conflict with the conservation of energy principle since the kinetic energy lost has not been destroyed but has merely changed form, into, for example, thermal energy (i.e. heat).

Two other possible collision situations should be mentioned. In some collisions the two bodies coalesce, in which case there is no bouncing off, and the collision is said to be *perfectly inelastic*. The other possibility was illustrated in example 4:2 above where the kinetic energy of the system increases as a result of the break-up of a body. The sudden increase of KE arises from the release of potential energy and such cases are referred to as *explosive* (or *superelastic*) collisions.

Example 4:6

A block of wood of mass 2.0 kg is suspended by long vertical threads and hangs in equilibrium. A bullet of mass 0.025 kg travelling horizontally with a velocity of 150 m/s strikes the block and becomes embedded in it. Calculate the horizontal velocity with which the block begins to move immediately after impact. Compare the kinetic energies before and after impact.

Solution The initial momentum of the system p_1 is that of the bullet, i.e.

$$p_1 = 0.025 \times 150 = 3.75 \text{ N s}$$

After impact, the block and bullet (of total mass $2 + 0.025$ kg) move with velocity v, and the momentum p_2 is

$$p_2 = 2.025 v \text{ N s}$$

Equating p_1 and p_2 on the basis of the conservation of momentum gives

$$2.025v = 3.75 \qquad v = 1.85 \text{ m/s}$$

The KE immediately before impact is

$$\tfrac{1}{2} \times 0.025 \times 150^2 = 281 \text{ J}$$

while the KE immediately after the collision is

$$\tfrac{1}{2} \times 2.025 \times 1.85^2 = 3.47 \text{ J}$$

We see then that in this perfectly inelastic collision nearly 99% of the kinetic energy is lost in the impact.

Exercises

Where necessary take the acceleration due to gravity $g = 9.8 \text{ m/s}^2$.

4.1 A block of mass 500 g slides down a frictionless plane of length 1 m inclined at 45° to the horizontal. Calculate its loss of potential energy. If the block started from rest at the top of the plane, determine its speed at the bottom.

4.2 An electron of mass 9.1×10^{-31} kg travelling at 10^6 m/s is brought to rest by an electric field in a distance of 5 cm. Calculate the loss of kinetic energy and the decelerating force on the electron.

4.3 A ball of mass 100 g is thrown vertically upwards at a speed of 25 m/s. If no energy is lost, determine the height it would reach. If the ball only rises to 25 m, calculate the work done against air resistance.

4.4 Water falls over a waterfall of height 30 m at a rate of 45 000 cubic metres per minute. Determine the power generated. (The density of water is 1000 kg/m.^{3})

4.5 A ball of mass 15 g strikes a vertical wall normally at a speed of 30 m/s and rebounds with a speed of 20 m/s. Calculate the change in momentum of the ball. If the time of contact between the ball and the wall is 1/40 sec calculate the average force exerted on the wall. What happens to the kinetic energy lost by the ball?

4.6 *a*) Explain what happens when a balloon filled with air is suddenly released so that air can freely escape from it.
b) A hovercraft can move over a water surface on a cushion of air. Explain how the craft is able to stay airborne.

4.7 A ball of mass 0.5 kg moving at 2 m/s overtakes and collides with a second ball of mass 1 kg moving at 0.5 m/s. If the 0.5 kg ball comes to rest give the new velocity of the 1 kg ball. Calculate the kinetic energy changes that occur.

4.8 A body of mass 200 g travelling at 0.2 m/s collides with a second body which is initially stationary. The two bodies stick together and the kinetic energy of the resulting motion is half that before collision. Find the final velocity and the mass of the second body.

4.9 In a radioactive beta-decay the unstable nucleus decays into a new nucleus, an electron and a neutrino. Explain the motion of the three products in the light of the principle of the conservation of momentum.

4.10 A pile-driver loses 49 kJ of potential energy in falling freely through a distance of 5 m onto a massive pile. The collision is completely inelastic and the velocity immediately after impact is 3 m/s. Calculate the kinetic energy lost by the pile-driver on impact.

4.11 In the situation described in exercise 4.10 above determine the impulse given to the pile by the impact of the driver.

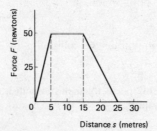

Fig. 4:10 Variation of accelerating force F with distance moved s.

4.12 The accelerating force on a body of mass 17.5 kg varies with distance s moved by the body as shown in fig. 4:10. Use the graph to calculate the total work done on the body. Hence determine the speed of the body after it has moved 25 m.

4.13 An electric motor drives a conveyor belt at a steady speed of 80 mm/s. If the force required to keep the belt moving is 2×10^4 N and only 80% of the power produced by the motor is utilised (i.e. the efficiency is 80%) calculate the power output of the motor.

4.14 Consider how the conservation of energy principle will apply to a spherical body rolling down an inclined plane. State what forms of kinetic energy the sphere will acquire in rolling.

4.15 A spanner of length 0.2 m is being used to tighten a nut. A force of 25 N at right angles to the spanner is applied at the end and the nut turns through 180°. Calculate the total work done.

4.16 A bar of moment of inertia 0.213 kg m^2 is rotating in a horizontal plane at an angular velocity of 9 rad/s. Determine the rotational kinetic energy and the angular momentum of the bar.

5 Vibrational Motion

This chapter considers **vibrational motion** which occurs for example in the swing of pendulums, in the thermal motion of atoms in a solid (see section 10:2), and in the movement of electric charges under an applied alternating voltage. Vibrational motion is important also because of its relationship to *wave motions*.

5:1 Simple Harmonic Motion

A convenient introductory example of vibrational motion is a loaded helical spring (fig. 5:1). If the load is pulled down from the equilibrium position and then released, it will undergo vibrational motion in a vertical plane. This motion has a number of important features.

In the first place, the motion is seen to repeat at a regular rate, that is it is a **cyclic** or **periodic motion**. The **period** is defined as the time for one complete cycle; that is, for example, the time to travel from B to A and back.

At points A and B (the extremes of the motion), the load must instantaneously come to rest since the direction of motion reverses. If the load is at A, then to start moving towards B it must be accelerated upwards (towards O). As the load approaches B, it will slow down again before momentarily stopping at B and then accelerating back towards A. It seems therefore that when the load is positioned between O and A, it experiences an upward acceleration, and when it is between O and B it has a downward acceleration. In other words the acceleration seems to always be towards O. Two further deductions follow from this analysis. First, the speed will be a maximum as the load passes the central point O and, second, since the direction of the acceleration reverses as the load passes through O, its magnitude must be zero at this point. These conclusions are set out in fig. 5:2.

The motion of the load on a helical spring is an example of **simple harmonic motion** (s.h.m.) as we will show later. In such a motion the acceleration at any instant is proportional to the displacement from equilibrium and is directed towards the equilibrium position. The acceleration can be written as dv/dt. This term was introduced in section 2:1 to denote the rate of change of velocity and its use here will serve as a reminder that the acceleration in s.h.m. is not uniform. The defining equation for s.h.m. can be written as

$$\frac{dv}{dt} \propto -x$$

where x is the displacement from the *origin* or equilibrium position. The negative sign indicates that the acceleration is always towards the equilibrium position.

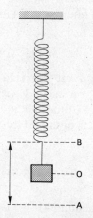

Fig. 5:1 Loaded helical spring vibrating between B and A. (O is the equilibrium position for the load.)

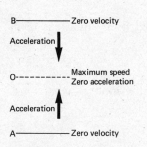

Fig. 5:2 Variation of velocity and acceleration in a vibrational motion.

The distance moved in a vibrational motion of any kind is specified by the **amplitude** a, which is the maximum displacement from the origin, so that in fig. 5:2,

$$a = OA = OB$$

The vibrating body therefore travels over a total distance $2a$.

5:2 Equations and Graphs of Simple Harmonic Motion

It is useful to be able to show graphically the variation of position, velocity and acceleration in s.h.m. To do this we need to know how these quantities change with time. This in turn requires us to solve the defining equation given above. This equation can be rewritten as

$$\frac{dv}{dt} = -\omega^2 x \tag{5:1}$$

where ω^2 has been introduced as a constant of proportionality whose physical meaning will be considered below. One solution of equation (5:1) is

$$x = a \sin \omega t \tag{5:2}$$

which gives the time variation of displacement x. The quantity a is a constant and, since the sine term can only vary between zero and ± 1, a will be the largest value of x. In other words, a is the amplitude of the motion. The velocity v (the rate of change of displacement) is obtained by differentiating equation (5:2) to give

$$\frac{dx}{dt} = v = a\omega \cos \omega t \tag{5:3}$$

Finally, if the value of x in equation (5:2) is substituted in equation (5:1) we obtain the expression for the acceleration:

$$\frac{dv}{dt} = -a\omega^2 \sin \omega t \tag{5:4}$$

Because these equations contain a sin (or a cos) term, they show that displacement, velocity and acceleration all **vary sinusoidally with time**, and the variations are shown in fig. 5:3.

Another useful equation which can be obtained from equations (5:2) and (5:3) is

$$v^2 = \omega^2(a^2 - x^2) \tag{5:5}$$

The maximum velocity occurs at the origin of the motion, where x is zero, so from equation (5:5) we find that

$$v_{max}^2 = a^2\omega^2$$

As a body performing s.h.m. moves between the points A and B in fig. 5:2, its velocity and hence its kinetic energy continuously change. The kinetic energy is maximum when the speed is maximum, so if the

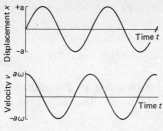

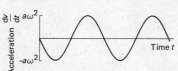

Fig. 5:3 Variations with time of displacement, velocity, and acceleration in s.h.m.

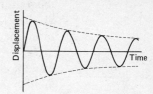

Fig. 5:4 Displacement/time graph for damped harmonic motion. The dashed curves show how the amplitude decreases with time.

vibrating system has mass m it follows from the expression above for v_{max} that its maximum kinetic energy, which occurs at the centre, is

$$KE_{max} = \tfrac{1}{2}ma^2\omega^2$$

In the absence of resistive forces, mechanical energy is conserved so that as the vibrating body moves away from the centre its KE decreases and its PE increases. In the case of a simple pendulum, its KE decreases as it swings either side of the equilibrium but, since the mass moves along the arc of a circle, it rises vertically, and consequently the gravitational PE increases correspondingly. If the PE is taken to be zero at the centre, the expression above for the maximum KE is also the total vibrational energy of the system.

Careful observation of the experimental situation shown in fig. 5:1 reveals that in practice the amplitude slowly decreases with time. This is because the vibrational energy of the system is slowly lost due to the effect of resistive forces. When such forces are present, a **damped harmonic vibration** results and the change of displacement with time is then as shown in fig. 5:4.

5:3 The Reference Circle

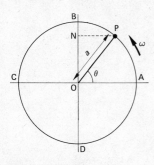

Fig. 5:5 Relating simple harmonic motion of point N to the reference circle.

The symbol ω in the above equations was introduced as a constant and we have yet to establish its identity. The quantity has however been met before in section 2:16 where we introduced the angular velocity. In fig. 5:5 a particle is shown moving round a circular path of radius a with constant angular velocity ω. If the particle starts from A at $t = 0$ and rotates anticlockwise to reach P after time t, then the angle turned through is

$$\theta = \omega t$$

The distance ON given by

$$ON = a \sin \theta = a \sin \omega t \qquad (5:6)$$

is called the *projection* of OP onto the y-axis, and N is located by dropping a perpendicular from P to the axis as shown. As P moves round the circle, N will move along the y-axis. As P goes once round the circle,

39

N travels from O to B, then to D, and finally back to O. The path of N is linear and its displacement from O at any time t is given by equation (5:6), which is the same form as equation (5:2). Thus the motion of N is simple harmonic and the path followed by the particle P is called the **reference circle**.

We see therefore that ω is the angular velocity of the corresponding circular motion. Both the circular motion of P and the simple harmonic motion of N will have the same period given by

$$T = \frac{2\pi}{\omega} = \frac{1}{f}$$

where f is called the **frequency** and is the number of complete vibrations of N each second. The unit of f is vibrations/second, and 1 vibration per second is 1 HERTZ, or 1 Hz.

It is sometimes useful to represent a simple harmonic motion by a line such as OP in fig. 5:5 (of length equal to the amplitude) which rotates at an angular velocity ω. OP is like a vector of constant magnitude, but changing direction, and the corresponding s.h.m. is the projection onto the axis. This is sometimes referred to as the *phasor representation* of s.h.m.

Example 5:1

A body performing s.h.m. of amplitude 0.05 m has a maximum acceleration of $20 \, \text{m/s}^2$. Calculate the frequency of the motion. Calculate also the maximum speed.

Solution   The maximum acceleration occurs at the extremes of the motion, so that from equation (5:1)

$$\left(\frac{dv}{dt}\right)_{max} = -\omega^2 a$$

hence (ignoring the minus sign which is concerned with the direction only)

$$20 = \omega^2 \times 0.05$$

$$\omega^2 = \frac{20}{0.05} \quad \text{giving} \quad \omega = 20 \, \text{rad/s}$$

Hence   Frequency $f = \dfrac{\omega}{2\pi} = 3.18 \, \text{Hz}$

The general equation for velocity (equation 5:5) is

$$v^2 = \omega^2(a^2 - x^2)$$

and since v is maximum when x is zero this gives

$$v_{max}^2 = 20^2 \times 0.05^2 \quad \text{or} \quad v_{max} = 20 \times 0.05 = 1.0 \, \text{m/s}$$

5:4 The Natural Frequency of a Loaded Spring

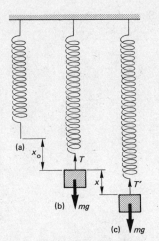

Fig. 5:6 (a) helical spring hanging vertically, (b) loaded at equilibrium, (c) displaced from equilibrium.

Fig. 5:6 shows a **helical spring** suspended from a fixed support. When a load of mass m is attached to the end, the spring stretches a distance x_0 to the equilibrium position shown in fig. 5:6(b). Experiment shows that the extension of a given spring is proportional to the load applied (a result known as Hooke's law) so that

$$mg \propto x_0 \quad \text{or} \quad mg = kx_0$$

where k is a constant called the spring constant, or the *stiffness*, and is the force required to produce unit extension.

At equilibrium the tension T in the spring is given by

$$T = mg = kx_0$$

If the load is then pulled down a distance x (fig. 5:6(c)), the tension increases to

$$T' = k(x_0 + x)$$

and the net force on the load is then $\quad T' - mg$ which on substituting gives

$$k(x_0 + x) - kx_0 = kx$$

When released the load will accelerate towards the equilibrium point and its equation of motion (from Newton's second law) is

$$-kx = m\frac{dv}{dt}$$

(the minus sign indicates a restoring force). The acceleration is therefore

$$\frac{dv}{dt} = -\left(\frac{k}{m}\right)x$$

and since k/m is constant

$$\frac{dv}{dt} \propto -x$$

which is the defining equation for s.h.m. So we conclude that the load undergoes s.h.m. with the constant term k/m equivalent to ω^2 in equation (5:1). The period of the motion will therefore be given by

$$T = \frac{2\pi}{\omega} = 2\pi \sqrt{\frac{m}{k}}$$

Once released, the spring vibrates freely so that its **natural frequency of vibration** is

$$f = \frac{1}{T} = \frac{1}{2\pi} \sqrt{\frac{k}{m}}$$

41

Example 5:2

A helical spring carries a mass of 100 g which performs s.h.m. of amplitude 80 mm. If the maximum speed is 0.6 m/s calculate the period and the stiffness of the spring.

Solution   To find the period we need to calculate ω. Since the maximum speed is given by

$$v_{max} = a\omega \quad \text{then} \quad \omega = \frac{0.6}{0.08} = 7.5 \text{ rad/s}$$

The period $T = 2\pi/\omega$ so that

$$T = \frac{2\pi}{7.5} = 0.84 \text{ s}$$

The period of a loaded spring is $T = 2\pi\sqrt{\dfrac{m}{k}}$

and squaring gives $T^2 = \dfrac{4\pi^2 m}{k}$

and on rearranging $k = \dfrac{4\pi^2 m}{T^2} = \dfrac{4\pi^2 \times 0.1}{(0.84)^2} = 5.6 \text{ N/m}$

5:5   The Natural Frequency of a Simple Pendulum

A **simple pendulum** consists of a small (point) mass m attached to a light string whose other end is fixed. When displaced and released, the mass swings along the arc of a circle in a vertical plane. We will assume that the angle θ through which the bob is displaced (as in fig. 5:7) is a small angle, which means that the path followed by the bob is nearly linear. Further, if x is the arc length, then for small angles

$$\theta \text{ (in radians)} = \frac{x}{l} \approx \sin\theta$$

When displaced, as in fig. 5:7, a net restoring force acts on the bob. We can equate the components of the forces along the string (since there is no motion in this direction) to give

$$T = mg\cos\theta$$

while normal to the line of the string the net force is

$$mg\sin\theta$$

which acts towards the equilibrium position (which is vertically below O). The equation of motion then becomes

$$-mg\sin\theta = m\frac{dv}{dt}$$

Substituting for $\sin\theta$ gives

$$-mg\frac{x}{l} = m\frac{dv}{dt}$$

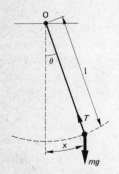

Fig. 5:7 Forces on a simple pendulum displaced from equilibrium.

so that $\dfrac{\mathrm{d}v}{\mathrm{d}t} = -\dfrac{g}{l}x$

When this is compared with equation (5:1) we see that the pendulum motion is simple harmonic, and that

$$\omega^2 = \frac{g}{l}$$

Hence we have

$$\text{Period } T = 2\pi\sqrt{\frac{l}{g}} \quad \text{and} \quad \text{Natural frequency } f = \frac{1}{2\pi}\sqrt{\frac{g}{l}}$$

Note that the frequency is independent of the mass of the bob, which may seem surprising but can be verified experimentally. Because the period can be changed simply by a change of length, pendulums are used in certain types of clock.

5:6 Forced Motion and Resonance

The systems whose motion we have investigated so far have vibrated as free systems.

A different situation will arise if the body is forced to vibrate, not at its natural frequency, but at a frequency imposed on it from outside. Consider the situation shown in fig. 5:8 where two bobs A and B are suspended by different lengths of thread from a taut string. The bob A is assumed to have much greater mass than B. This situation can be reproduced by hanging 50 g and 5 g masses between two retort stands. When mass A is displaced in a plane perpendicular to the diagram and released, it acts as the driver for bob B and forces it into motion. After a short time a steady state is reached and it is found that both bobs perform s.h.m. with the same period. Measurement of the period shows it to be the value corresponding to a pendulum of length l_1. Bob B is therefore forced to vibrate at the frequency of the driver rather than its own natural frequency.

Further investigation shows that the amplitude of bob B changes as the length of its supporting thread changes. The amplitude increases as l_2 gets nearer to l_1 and becomes maximum when l_2 equals l_1. In this situation the driver frequency is equal to the natural frequency. This amplitude maximum is called **resonance** and occurs, in systems where there is little damping, when the system is driven at its natural frequency. Fig. 5:9 shows the general form of the change in amplitude as the driver frequency changes.

Some startling consequences can result from resonance. In 1940 a suspension bridge across the Tacoma Narrows in the United States broke up as a result of being set into vibration at its resonant frequency by wind gusts during a gale. A wine glass also has a natural frequency and, if a singer holds a note of the same frequency, the glass will vibrate and shatter.

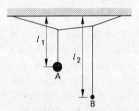

Fig. 5:8 Coupled system of two bobs to illustrate forced motion.

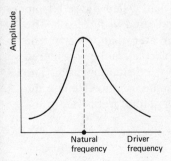

Fig. 5:9 Variation of amplitude with driver frequency for forced motion, showing the amplitude peak at resonance.

5:7 Further Examples of Resonance

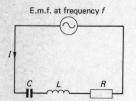

Fig. 5:10 Series a.c. circuit with capacitance C, inductance L and resistance R.

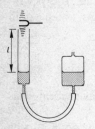

Fig. 5:11 Resonance tube. The air column is excited by a vibrating tuning fork.

Fig. 5:12 Schematic representation of the HCl molecule as an electric dipole.

The examples of resonance given above all involved mechanical vibrations. However resonance is not restricted to such cases and we may briefly consider three other examples.

1 Fig. 5:10 shows a source of *alternating emf supplying power to a series circuit* consisting of a capacitor C, inductor L and resistor R. The size of the alternating current through the circuit depends on the frequency of the supply. Maximum current flows at the resonant frequency which can be shown to be

$$f = \frac{1}{2\pi\sqrt{[LC]}}$$

where L is the inductance and C the capacitance in the circuit.

2 Our second example comes from *acoustics*. Fig. 5:11 shows an air column whose length can be varied by means of a water reservoir. A vibrating tuning fork held over the end of the tube sets the air column into vibration. Generally these vibrations are not of sufficient amplitude to be heard. However as the length of the air column varies, it is found that there are certain lengths at which the air vibration increases sufficiently to be audible. An air column has a number of natural frequencies, which depend on the column length, and when these equal the frequency of the fork the air column resonates.

3 The final example shows resonance at the *molecular level*. Common salt (NaCl) forms an ionic crystal in which sodium ions (Na^+) and chloride ions (Cl^-) are held in the lattice by electrostatic forces. If the crystal is exposed to electromagnetic radiation, the electric field of the wave sets the ions into vibration which is maximum when the frequency of the radiation is equal to the natural frequency of vibration of the ions. The crystal then shows maximum absorption of energy. For NaCl the natural frequency lies in the infra-red part of the spectrum around 5×10^{12} Hz.

In an HCl molecule the electron from the hydrogen atom is shared rather than transferred so that the bonding is covalent. The distribution of charge within the molecule however is such that, when viewed from outside, the molecule shows the dipole form indicated in fig. 5:12. The mass of the chlorine atom is considerably greater than that of the hydrogen atom, so that the vibrations of the molecule can be considered to be the motion of a hydrogen atom bound to a fixed chlorine atom. The equation for the natural frequency of a spring can be used to estimate the force constant for the HCl molecule, which has a natural frequency in the infra-red at 9×10^{13} Hz.

Using $f = \frac{1}{2\pi} \sqrt{\frac{k}{m}}$

$$9 \times 10^{13} = \frac{1}{2\pi} \sqrt{\left[\frac{k}{1.67 \times 10^{-27}}\right]}$$

giving $k = (9 \times 10^{13})^2 \times 4\pi^2 \times 1.67 \times 10^{-27} = 534\,\text{N/m}$

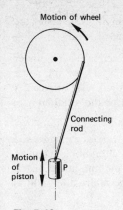

Motion of wheel

Connecting
rod

Motion
of
piston

P

Fig. 5:13

Measurement of the resonant frequency of molecules can give us information about the strength of the interatomic forces.

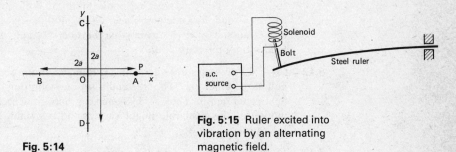

Fig. 5:14

Fig. 5:15 Ruler excited into vibration by an alternating magnetic field.

Exercises

Where necessary take the acceleration due to gravity $g = 9.8 \, \text{m/s}^2$.

5.1 A spring of force constant $10 \, \text{N/m}$ carries a load of $60 \, \text{g}$. Give its natural frequency of vibration.

5.2 If the amplitude of the displacement in exercise 5.1 is $5 \, \text{cm}$ find the maximum acceleration of the load.

5.3 A mass attached to the free end of a light rod swings as a simple pendulum with a period of $2.1 \, \text{sec}$. If the rod is shortened by $0.2 \, \text{m}$ calculate the new period.

5.4 A piston P is connected by a long rod to the rim of a wheel of diameter $0.6 \, \text{m}$ which rotates about its centre as shown in fig. 5:13. The piston is confined to motion along its axis as shown. If the wheel rotates at $600 \, \text{rpm}$ determine the amplitude and maximum velocity of the piston motion.

5.5 The piston of an engine moves over a total distance of $10 \, \text{cm}$. If its motion is assumed to be simple harmonic with a frequency of $50 \, \text{Hz}$ find the speed of the piston at a point $2.5 \, \text{cm}$ from the centre of the motion.

5.6 In a particular solid the force constant between the ions is $60 \, \text{N/m}$. If the mass of the ions is $10^{-25} \, \text{kg}$ and it is assumed that they vibrate with s.h.m. calculate the frequency of the vibration.

5.7 A pendulum bob of mass $200 \, \text{g}$ swings with a period of $2 \, \text{s}$ and amplitude $10 \, \text{cm}$. Determine the maximum vibrational kinetic energy. Determine where in the path the potential energy of the pendulum becomes maximum.

5.8 If the pendulum in exercise 5.7 is subjected to a damping force so that at some later time half of its kinetic energy has been lost calculate the new amplitude. State whether the maximum potential energy will also be reduced.

5.9 Sketch a graph showing the variation of kinetic energy with position between the extremes of a simple harmonic motion. Explain how the total energy changes with position.

5.10 The displacement of a certain body performing a simple harmonic vibration is given by the equation

$$x = 0.1 \sin 4\pi t$$

where distances are in metres and time in seconds. Calculate the angular velocity of the reference circle motion and the period of the s.h.m. Calculate the displacement after 0.1 sec. (Watch the units used for the angle.)

5.11   Fig. 5:14 shows the instantaneous position of a particle P which is moving under the action of two simple harmonic motions, both centred at O, of the same frequency and the same amplitude *a*. One motion acts along the *x*-axis between A and B and the other along the *y*-axis between C and D. Describe the resultant path followed by the particle P.

5.12   Fig. 5:15 shows a steel ruler which is clamped at one end and carries a magnetised bolt at the other end. The free end is set into vibration by an alternating magnetic field generated in a solenoid. Sketch a graph showing how the amplitude of the motion of the steel rule might vary as the frequency of the alternating field is increased.

5.13   Distinguish between free vibrations and forced vibrations. Explain why certain parts of a car, such as the door panels, vibrate strongly at certain engine speeds.

5.14   Obtain equation (5:5) from equations (5:2) and (5:3).

6 Waves and their Basic Properties

Wave phenomena are very widespread in the physical world. We use sound waves in speech communication while radio and television signals are forms of electromagnetic waves. An even more familiar example is the wave motion on the surface of the sea. In describing such wave motions the physicist looks for the underlying similarities between the many different types of waves.

What is a wave motion? A simple example may help to provide an answer. A stone dropped into the middle of a pool of water produces ripples on the surface which spread outwards from the point of impact. Some of the energy of the stone has been used to produce a disturbance on the water surface which travels away from the source. We can think of a wave motion as the transfer of energy by means of a travelling disturbance. In the example here there is, of course, no bulk motion of the water. Instead the surface is displaced vertically as the ripples travel across the surface. This vertical motion could easily be seen if a cork floated on the surface.

6:1 Describing a Wave Motion

A wave motion can be represented by the use of graphs. Imagine a long string (with one end fixed) held taut and horizontal as shown in fig. 6:1(a). If the free end is moved rapidly up and down, a disturbance travels down the string. If this motion could be frozen (perhaps by taking a photograph), the string would appear as shown, with the diagram representing the **displacement** (y) of different positions along the string (x). This sort of diagram is a y/x plot. If the motion of a particular point along the string, such as A, is carefully observed it is found to vibrate vertically. If this motion is displayed graphically it would look like fig. 6:1(b), which shows the variation with **time** of the displacement of some given point in the medium. (This graph is called a y/t plot.) Both y/x and y/t plots can represent a wave motion.

Fig. 6:1 (a) A y/x plot showing the disturbance of a medium "frozen" at a given time; (b) a y/t plot showing the motion of point A in the medium as the disturbance passes.

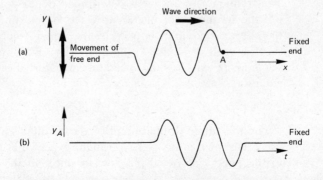

A y/t plot is obtained in the arrangement shown in fig. 6:2 where the sound wave emitted by a loudspeaker is detected at a microphone and displayed on a cathode ray oscilloscope (CRO). The sound wave is produced in the air by vibrations of the cone of the loudspeaker. In the simplest case where the vibration is simple harmonic, a sinusoidal waveform is produced.

The disturbance set up in a string, as in fig. 6:1, is found to travel at a speed which depends on the tautness and thickness of the string. The expression for the speed v is

$$v = \sqrt{\frac{T}{\mu}}$$

Fig. 6:3 (a) y/x plot and (b) y/t plot showing basic wave quantities.

where T is the tension in the string and μ is the mass per unit length. This illustrates a general rule that the characteristics of the medium determine the speed with which wave energy can travel through the medium.

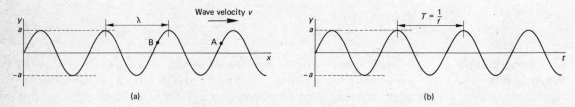

(a) (b)

Fig. 6:3(a) shows a general y/x plot for a continuous wave. The distance λ between the peaks is called the **wavelength**. As the disturbance travels, a series of peaks pass a given point. The number passing each second is called the **frequency** f. Fig. 6:3(b) shows the motion at a particular point. The time interval between the peaks is the period T and we saw in section 5:3 that

$$T = \frac{1}{f}$$

The frequency of a wave is determined by the source. For example in fig. 6:2 the frequency of the sound wave is determined by the frequency of the alternating e.m.f. supplied by the signal generator. If f waves pass per second, each of length λ, the distance moved by the disturbance in one second is $f\lambda$, which is the speed of the disturbance, so we have the general equation

$$v = f\lambda \qquad \qquad (6:1)$$

The maximum displacement of the medium gives the **amplitude** of the wave as shown in fig. 6:3. We might expect that this amplitude is related to the energy carried by the wave since, in the case of the taut string, more vigorous movement of the free end, i.e. putting more energy in, results in a larger disturbance. In section 5:2 we found that the energy of a vibrating system was proportional to the square of the amplitude a. The same relationship holds for a wave motion so that, for **energy** E,

$$E \propto a^2$$

We can also define the **intensity** I of a wave as the energy flowing per second through unit area normal to the wave. The intensity also depends on the square of the amplitude.

Example 6:1

Radio Chiltern broadcasts on 362 m on the medium wave and 97.5 MHz v.h.f. If the speed of radio waves is 3.0×10^8 m/s determine the frequency of the medium wave signal and the wavelength of the v.h.f. signal.

Solution   In each case we can use the general equation (6:1)

$$v = f\lambda$$

so that for the medium wave signal of wavelength 362 m

$$3.0 \times 10^8 = 362 \times f$$

$$f = \frac{3 \times 10^8}{362} = 8.29 \times 10^5 \text{ Hz or } 829 \text{ kHz}$$

For the v.h.f. signal of frequency 97.5×10^6 Hz we have

$$3.0 \times 10^8 = 97.5 \times 10^6 \lambda$$

$$\lambda = \frac{3 \times 10^8}{97.5 \times 10^6} = 3.08 \text{ m}$$

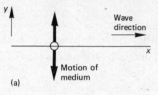

(a)

(b)

Fig. 6:4 Direction of vibration in (a) transverse wave, (b) longitudinal wave.

6:2 Transverse and Longitudinal Waves

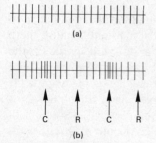

(a)

(b)

C   R   C   R

Fig. 6:5 (a) Undisturbed layers of air; (b) layers displaced by a longitudinal wave producing compressions (C) and rarefactions (R).

When a disturbance travels down a stretched string, the medium undergoes vibration in a direction at right angles to the direction of travel as shown in fig. 6:4(a). Such a wave motion is said to be **transverse**. Surface water waves are also transverse since a cork on the surface bobs up and down as the wave passes. However the motion in a sound wave is different. Here the air vibrates in the same direction as the wave, as illustrated in fig. 6:4(b), and the wave is described as **longitudinal**. It is however still possible to represent a longitudinal wave by a y/x or y/t plot provided we remember that the displacement (y) takes place along the wave direction (the x direction).

Fig. 6:5 illustrates a sound wave travelling through air. Initially the layers of air are equally spaced, corresponding to uniform pressure through the medium. When the sound wave passes, the air layers are displaced. At any given time some layers will move forward while others move back, so that the layers appear as in fig. 6:5(b). At some points (C), called *compressions*, the layers are pushed together so that the pressure increases slightly, while at other points (R), called *rarefactions*, the layers are further apart and the pressure is reduced. The propagation of a sound

49

wave therefore leads to small pressure changes in the medium, and for this reason longitudinal waves are also called **compressional** waves. In a domestic living room with the radio on, the pressure change is unlikely to exceed about 0.02 Pa. Since atmospheric pressure is approximately 1×10^5 Pa the pressure change is less than one millionth of an atmosphere. [Pressure and its unit, the Pascal (Pa), are dealt with in section 11:3.]

Fig. 6:6 (a) Transverse and (b) longitudinal disturbance in a slinky spring.

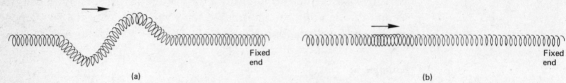

(a)

(b)

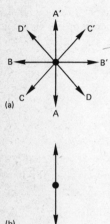

(a)

(b)

Fig. 6:7 View along a stretched string: (a) shows some possible directions of transverse vibration; (b) shows vibration restricted to one direction in a plane polarized wave.

Longitudinal and transverse motions can be illustrated using a slinky spring. The spring is laid out on a long table with one end fixed. If the free end is moved normal to the line of the spring, a transverse disturbance is set up (fig. 6:6(a)), and compressing the windings near the free end and then releasing will produce a longitudinal disturbance as in fig. 6:6(b).

In the case of the stretched string in fig. 6:1, the wave is set up by moving the free end vertically. The free end could equally be moved horizontally, or along CC' or DD' (fig. 6:7(a)), to produce a transverse disturbance. In practice a wave such as the light emitted by the sun or by a filament lamp may consist of vibrations in many directions in this plane normal to the wave direction. Such a wave is said to be unpolarized.

In other cases vibrations may be restricted to one particular direction and the wave is said to be **plane-polarized**. Such a wave is represented in fig. 6:7(b) with the plane of vibration vertical. The wave in fig. 6:1 is plane-polarized since the free end is moved in just one of the possible directions shown in fig. 6:7(a). We shall consider polarized light in more detail in chapter 7.

In a longitudinal wave the direction of vibration is fixed by the direction of travel of the wave and the medium can only move in this one direction as fig. 6:4(b) shows. Hence the distinction between polarized and unpolarized waves does not apply to longitudinal waves.

6:3 Mechanical and Electromagnetic Waves

A **mechanical wave** is a wave which requires a **medium** through which to travel. The energy of the wave travels through the medium because there is some form of mechanical coupling between different parts of the medium. Without a medium energy cannot travel, so, for example, sound waves will not propagate through a vacuum.

Electromagnetic waves such as light, however, can travel across a **vacuum**. The disturbance in these waves is not the physical movement of a medium such as a string but the variation of electric and magnetic fields which are set up by the wave source. These electromagnetic waves, which can cover a wide frequency range, will be considered in chapter 8.

The fundamentally different natures of mechanical and electromagnetic waves is shown by the different speeds of the waves. All electromagnetic waves have the same speed of 3.0×10^8 m/s in a vacuum. The presence of a physical medium reduces the speed so that, for example, light travels in glass at about 2×10^8 m/s. Mechanical waves have much smaller speeds. For sound waves in air the speed is related to the pressure p and density ρ by

$$v = \sqrt{\frac{\gamma p}{\rho}}$$

where γ is a constant which, for air, has a value of 1.4. (The constant will be defined in section 11:6.) In the atmosphere at 20°C the speed of sound is about 340 m/s.

6:4 Fronts and Rays

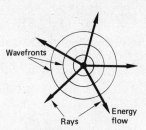

Wavefronts

Energy flow

Rays

Fig. 6:8 Wavefronts and rays from a point source.

When a stone is dropped into a pond, ripples are seen to spread out from the point of impact. These expanding ripples show energy moving away from the source. The ripples form a series of **wavefronts** which spread outwards at the speed of the wave. The wavefronts on the surface of a pond will be circular, while those associated with a *point source* such as a small lamp will be spherical. In general the wavefronts join points which have the same displacement at a given time.

Fig. 6.8 shows the wave emission from a point source represented by a series of concentric circles corresponding to the peaks of the wave. The circles are therefore one wavelength apart. The energy which travels away from the source moves along the radial lines shown in the diagram. These **rays** will be at right angles to the wavefront at any point in the medium.

6:5 Reflection and Refraction

The two basic wave properties of reflection and refraction can be considered using the idea of wavefronts. Fig. 6:9 shows a plane wavefront AB approaching a reflecting surface at an *angle of incidence i*. (This angle is measured between the wave direction, or ray direction, and the normal to

Fig. 6:9 Reflection of a plane wavefront at a reflecting boundary, to show that $i = r$.

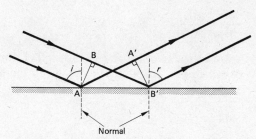

Normal

the boundary.) The wavefront advances with time and at a later instant reaches position A'B'. Its direction after reflection is measured by the angle r, called the *angle of reflection*. If there is no change of speed of the wave then

$$AA' = BB'$$

The triangles ABB' and B'A'A have AB' as a common side and both contain a right angle so that the two triangles are congruent. Hence

$$\text{angle BAB}' = \text{angle A}'\text{B}'\text{A}$$

i.e. angle of incidence i = angle of reflection r

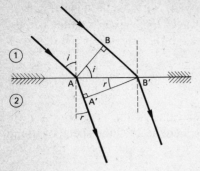

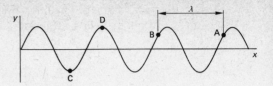

Fig. 6:11 *y/x* plot illustrating idea of phase difference. A and B are in phase, C and D are in antiphase.

Fig. 6:10 Refraction of a plane wavefront at a boundary. The normals are shown dashed.

This basic **law of reflection** follows directly from the fact that there is no change of speed on reflection.

When a wave passes from one medium to another, its speed changes and refraction occurs as shown in fig. 6:10. Here a wavefront AB is incident at angle i on a plane boundary between media 1 and 2. A'B' is the wavefront position at a later time. The angle r made by the front in the second medium is called the *angle of refraction*. If the speeds in the media are v_1 and v_2, then in the same time interval t, A moves to A' and B to B' and

$$AA' = v_2 t \qquad BB' = v_1 t$$

In triangle ABB' we have $\quad \sin i = \dfrac{BB'}{AB'}$

while triangle AA'B' gives $\quad \sin r = \dfrac{AA'}{AB'}$

Dividing these gives the ratio

$$\frac{\sin i}{\sin r} = \frac{BB'}{AA'} = \frac{v_1 t}{v_2 t} = \frac{v_1}{v_2}$$

The ratio of the sines of the angles is therefore equal to the ratio of the speeds which is a constant for two given media. This relationship

$$\frac{\sin i}{\sin r} = \text{constant}$$

is the basic **law of refraction** and is called Snell's law. The constant is called the *refractive index n* for the two media and is determined by the speeds. In the case shown in fig. 6:10 the wavefront direction changes so that the wave travels nearer to the normal, i.e.

$$\sin i > \sin r \quad \text{hence} \quad v_1 > v_2$$

This is the situation when light travels from air into glass or water.

We have seen here that refraction, or the bending of a wave at a boundary, is the result of a change of speed at the boundary.

Example 6:2

Sound waves travelling in air strike an air/water boundary with angle of incidence

10°. If the sound wave entering the water has an angle of refraction of 46° determine the speed of sound in water. Take the speed in air to be 340 m/s.

Solution Since $r > i$ the wave is bent away from the normal and so the speed in water will be greater than in air. Applying the law of refraction gives

$$\frac{\sin i}{\sin r} = \frac{v_{air}}{v_{water}} \quad \text{i.e.} \quad \frac{\sin 10°}{\sin 46°} = \frac{340}{v_{water}}$$

$$v_{water} = \frac{340 \sin 46°}{\sin 10°} = \frac{340 \times 0.719}{0.174} = 1408 \text{ m/s}$$

6:6 Phase Difference

In the wave motion shown in fig. 6:11 the points A and B (one wavelength apart) have the same displacement at a given time. As the wave travels, these points in the medium will both undergo s.h.m. but will continue to have the same instantaneous displacements. In other words the two points move in step. The movement of these points is said to be **in phase**.

The points C and D show a **phase difference** since, at the instant shown, D has maximum positive displacement while C has maximum negative displacement. Phase difference is usually expressed as an angle in radians. Points A and B are one wavelength, or one cycle, apart. Since one cycle corresponds to rotation through 2π radians, points C and D, which are half a wavelength apart, will have a phase difference of π radians. In general if two points are distance x apart the phase difference between them is

$$\phi = \frac{2\pi x}{\lambda}$$

Fig. 6:12 Superposition of waves: (a) two waves in phase giving constructive interference; (b) two waves with π phase difference giving destructive interference.

This idea of phase difference can also be used where two waves, perhaps from different sources, arrive at the same point. Fig. 6:12 shows two waves A and B, at the same point, and y_A and y_B are the disturbances due to each. In fig. 6:12(a) both waves produce displacement in the same direction at any given time and the two disturbances are in phase. The effect of the two waves is to give a net displacement which is

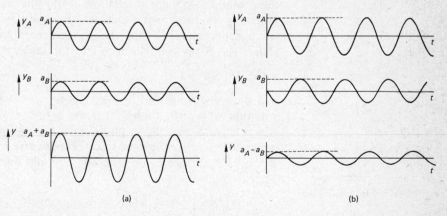

(a)   (b)

the sum of the separate displacements:

$$y = y_A + y_B$$

and, with waves in phase, the resultant amplitude a is

$$a = a_A + a_B$$

In fig. 6.12(b) the two waves show a phase difference of π radians. Wave A produces positive displacement when wave B displaces negatively. When the waves are added, or superposed, the net displacement will be reduced and, since the waves are always acting in opposition, the amplitude is reduced to

$$a = a_A - a_B$$

The result of adding two disturbances is therefore not only a matter of the amplitudes of each but depends also on the phase relationship between them. Fig. 6:12(a) shows a case of *constructive interference*, where the waves produce a larger disturbance, while fig. 6:12(b) shows *destructive interference* with the waves tending to cancel out. Destructive interference arises in cases where the phase difference is such that the resultant amplitude is reduced.

6:7 Superposition and Interference

When two waves arrive at a point their individual displacements can be added to give the resultant disturbance. This is the *principle of superposition of waves*. Note that it is the displacements which add rather than the energies. This principle can be extended to cases where a number of waves are involved. The net disturbance y of a point in the medium is found by summing the displacements due to each individual wave, i.e.

$$y = y_A + y_B + y_C + \cdots$$

In interference we are usually dealing with just two waves.

If two waves from fixed sources (S_1 and S_2) meet at a point (fig. 6:13), the resultant amplitude will depend on the individual amplitudes and the phase difference between the waves. The phase difference will depend on the position of point P in relation to the sources. If P is equidistant from both sources, those wavefronts which leave the sources at the same time will reach the point P together, with zero phase difference. But if one source (say S_2) is some distance x further away there is a path difference

$$S_2P - S_1P = x$$

and the wavefront from S_2 takes longer to reach P. Fig. 6:13 shows the situation where the path difference is half a wavelength so that waves which leave S_1 and S_2 in phase have a phase difference of π at P and destructive interference will result. **Destructive interference** will also arise if x is 1.5λ, 2.5λ, etc., or generally any **odd number of half wavelengths.** That is

$$x = (2n + 1)\tfrac{1}{2}\lambda \qquad \text{where } n \text{ is an integer}$$

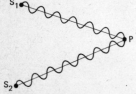

Fig. 6:13 A path difference from S_1 and S_2 to P results in a phase difference between the waves arriving at P.

Fig. 6:14 One form of ripple tank.

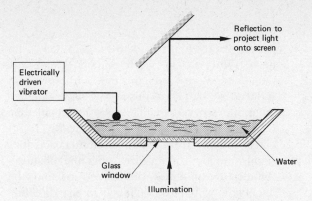

Constructive interference will occur if the path difference x is one wavelength or a **whole number of wavelengths**, i.e.

$x = n\lambda$ where n is an integer

Effects which result from superposition can readily be seen on a ripple tank, one form of which is shown in fig. 6:14. The tank consists of a shallow pool of water which is illuminated from below. Transverse waves are set up on the surface by an electrically driven vibrator which just touches the surface. The motion of the ripples across the surface can be frozen by rotating a disc with holes cut in it in front of the light source. The ripples are only illuminated at certain times and by varying the rate of rotation of the disc, the ripples can be made to appear stationary and be viewed on a screen.

The superposition of two waves can be seen on a ripple tank. The result is shown in fig. 6:15. It can be seen that the surface is crossed by a series of lines along which the ripples have cancelled out. These lines join points where the path differences are an odd number of half wavelengths, i.e. destructive interference results. Between these lines the waves interfere constructively and the resultant amplitude is maximum.

Fig. 6:15 Form of the two-source interference pattern on a ripple tank.

Example 6:3

Two loudspeakers L_1 and L_2 driven by the same signal generator are positioned 18 cm apart and a microphone M is located 80 cm in front of one speaker as shown in fig. 6:16. If the wavelength of the sound waves is 12 cm calculate the phase difference between the two waves arriving at the microphone.

Solution   The path difference $x = L_1M - L_2M$

Since L_1L_2M is a right-angle triangle we can write

$$(L_1M)^2 = (L_1L_2)^2 + (L_2M)^2 = 18^2 + 80^2 = 324 + 6400 = 6724$$

$$L_1M = 82 \text{ cm}$$

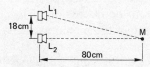

Fig. 6:16 Position of loud-speakers and microphone.

Hence the path difference $x = 82 - 80 = 2$ cm.

Using $\phi = \dfrac{2\pi x}{\lambda}$ $\phi = \dfrac{2\pi \times 2}{12} = \dfrac{\pi}{3}$ radians (or 60°)

6:8 Diffraction at a Slit

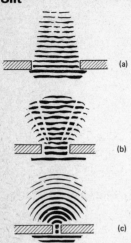

Fig. 6:17 Plane wavefront incident on a narrow slit showing diffraction effect as width decreases.

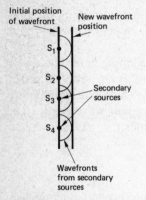

Fig. 6:18 Progression of a plane wavefront using the idea of secondary sources.

Diffraction of water waves may be shown on a ripple tank by placing a slit in front of a plane wave source and looking at the effect of reducing the slit width. Fig. 6:17 shows the effect at three different slit widths. In fig. 6:17(a) the slit width is much greater than the wavelength λ, and the plane waves pass through with little modification. If the width is reduced so that it is just a few wavelengths (about 3λ), the wavefront emerging from the slit shows noticeable curvature at the edges (fig. 6:17(b)). Finally when the slit width is about one wavelength, the emergent wavefront is almost circular and the slit appears to behave as a small source. With narrow slits the obstacle in the path of the wave significantly modifies the wavefront and the consequent direction of energy flow. It is this change in the direction of the wave energy flow which is called **diffraction** and fig. 6:17 shows it to be most significant when the slit size is comparable to the wavelength.

A further effect can be seen in fig. 6:17(b). At certain angles there appears to be very little energy flow. It is beyond the scope of this book to attempt a full explanation of this but we can indicate why it occurs. Diffraction effects can be considered to result from the superposition of a large number of small wavelets. A wavefront such as that shown in fig. 6:18 can be treated as though it were made up of a large number of wave sources, called *secondary sources*, each acting like a point source. Four such sources are shown along the plane wavefront. Each secondary source produces a circular wavefront and the position of the plane wavefront at a later time is found by drawing the envelope to these secondary wavefronts.

The slit in fig. 6:17 can be treated as a series of secondary sources and the intensity in any direction on the emergent wavefront can be found by adding the disturbances due to each secondary source. If this is done it is found that at certain angles the disturbances superpose in such a way that the resultant is zero as seen in fig. 6:17(b).

This effect can also be seen with light. A convenient source to use is a laser which gives a narrow beam of intense monochromatic (that is a single wavelength) radiation. This light passes through a narrow slit and the resulting diffraction pattern is displayed on a screen (fig. 6:19). The intensity variation across the screen is sketched in the diagram and also shown in the photograph (fig. 6:20). The laser acts like a point source and, with the slit vertical, a series of horizontally spaced secondary maxima are observed with zero intensity at certain angles between them.

6:9 Standing Waves

In all the wave situations we have referred to so far a source has set up a disturbance which then travels through the medium. Different points in the medium perform s.h.m. at different phases (see the discussion of fig. 6:11 in section 6:6), and energy is transferred through the medium. Such a wave motion is said to be **progressive**.

A different situation will result where two waves of the same frequency travel in opposite directions. The two waves interfere and, by applying the principle of superposition, we can calculate the displacement at

Fig. 6:19 Demonstration of single slit diffraction using laser light.

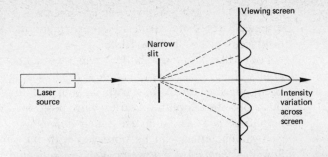

Fig. 6:20 Photograph of intensity variation across the screen in fig. 6:19.

Fig. 6:21 Experiment to show a standing wave in a stretched string.

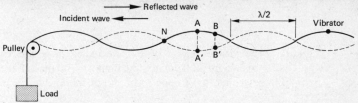

different points in the medium. It is found that the motion of the medium is quite different to the progressive wave case and there is no net transfer of energy. The superposition produces a **standing wave**.

Fig. 6:21 shows an experimental arrangement to illustrate the form of the standing wave motion. A horizontal stretched string passes over a pulley and is held under tension by an applied load. The other end is attached to a vibrator which sets up a transverse disturbance which travels down the string and is reflected at the pulley. At any point along the string there are therefore two waves moving in opposite directions.

Superposition of these gives the effect shown. The solid line indicates the displacement of the string at a particular time and the dashed line the displacement half a period later.

Points like N in fig. 6:21 do not move at all (and are called **nodes**) while points like A vibrate between A and A'. Point B vibrates between B and B' in phase with A but with smaller amplitude. In a standing wave therefore the amplitude varies with position. Points like A where the amplitude is a maximum are called *antinodes*.

A similar effect can be observed with the set-up shown in fig. 6:22. Sound waves from a loudspeaker are reflected at a metal plate so the microphone receives two signals arriving from opposite directions. The cathode ray oscilloscope displays the resultant intensity at the microphone. As the microphone moves towards the reflector, the resultant signal is found to rise and fall as the microphone passes through successive nodes and antinodes. Theory shows that the separation of the nodes is $\lambda/2$ as shown in fig. 6:22 and, if the frequency of the vibrator or signal generator is known, we can use the equation

Fig. 6:22 Experimental arrangement to show a standing sound wave.

$$v = f\lambda$$

to determine the velocity of the waves.

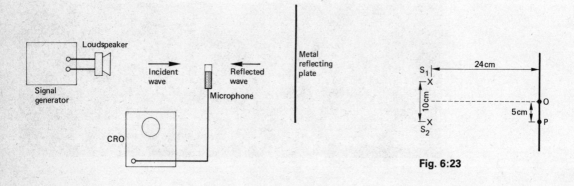

Fig. 6:23

Exercises

6.1 The visible region of the electromagnetic spectrum runs from violet light of wavelength 400 nm to red light at 700 nm. If the speed of light is 3×10^8 m/s determine the frequency range of the visible region.

6.2 A steel wire of mass per unit length 8×10^{-3} kg/m is fixed at one end. The wire passes over a frictionless pulley and the free end carries a load of 5 kg. Determine the tension in the wire and the speed of a transverse wave travelling down the wire. (Take the acceleration due to gravity as 9.8 m/s².)

6.3 During a storm it is observed that the time interval between seeing a lightning flash and hearing the clap of thunder is 6 s. If the speed of sound is 340 m/s determine how far away the storm is from the observer.

6.4 Two surface water waves of the same amplitude and frequency arrive at a point on the surface in phase. Give the ratio of the intensity of the resultant wave to the intensity of the individual waves.

6.5 Explain what you would expect to happen to the speed of sound in air as the temperature of the air increases. (You could consider what factors determine the speed and how these would be affected by a temperature change.)

6.6 A sound wave travels through carbon dioxide gas which has a density of 1.97 kg/m^3 at normal pressure and temperature (1.013×10^5 Pa and 273 K). If γ for carbon dioxide is 1.3 determine the speed of the wave at 273 K.

6.7 Indicate which wave properties change, and which remain unchanged, when a wave crosses the boundary from one medium to another.

6.8 Light rays passing from air to water are bent towards the normal. If the refractive index between air and water is 1.33 determine the speed of light in water. (Take the speed in air to be $3 \times 10^8 \text{ m/s}$.)

6.9 Two loudspeakers are driven by the same signal generator at a frequency of 1000 Hz so that they produce waves which are in phase. A microphone is positioned so that it is 2.5 m from one speaker and 2.6 m from the other. If the speed of sound is 333 m/s find the path difference as a fraction of a wavelength and determine the phase difference between the two waves at the microphone.

6.10 Fig. 6:23 shows two sources (S_1 and S_2) of ultrasonic waves of wavelength 1 cm. The waves arriving at point O, which is equidistant from the two sources, are in phase and the intensity of the resultant wave is maximum. State whether the intensity at point P, positioned as shown, will be maximum or minimum.

6.11 A microphone placed between a loudspeaker and reflector as shown in fig. 6:22 is initially at a point where the intensity is a minimum. In moving 28 cm towards the loudspeaker the intensity at the microphone goes through three minima. If the frequency of the sound wave is 1800 Hz calculate the speed of sound.

6.12 A steel wire of mass per unit length 2×10^{-3} kg/m is stretched between two fixed bridges 1 m apart. The wire passes between the pole pieces of a magnet placed at its centre so that, when an alternating current of frequency 50 Hz is passed through the wire, it is set into vibration with a node at each end and an antinode at the centre. Find the speed of the transverse wave in the wire and the tension.

7 Aspects of Optics

In this chapter we will use the concept of rays to explain the behaviour of light in mirrors, prisms and lenses, and then use a wave treatment to discuss interference, diffraction and polarization.

7:1 Plane Mirror Reflection

As an example of the reflection of light consider the behaviour of plane mirrors. At a reflecting boundary, the angle of incidence i is equal to the angle of reflection r (section 6:5), and this result can be applied to image formation. In fig. 7:1 two rays from an object O undergo reflection and, to a viewer, appear to have come from point I, behind the mirror, which is the image position. No light actually travels to this image, so it is virtual rather than real.

It can be shown from simple geometry that the image distance v from the mirror is equal to the object distance u. Hence the image is the same size as the object. A finite object when viewed in the mirror will appear *laterally inverted* as shown in fig. 7:2.

Plane mirror reflection is applied in the optical lever to measure small rotations (for example in a sensitive moving coil galvanometer). Fig. 7:3 shows a small mirror M which reflects a narrow beam onto a screen S. If the mirror rotates through angle θ to M', the reflected ray rotates through twice this angle and the result is a greater movement of the spot on the screen.

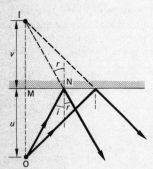

Fig. 7:1 Image formation in a plane mirror.

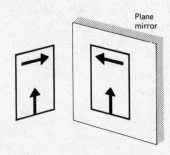

Fig. 7:2 Example of lateral inversion in a plane mirror.

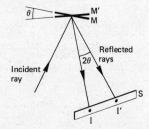

Fig. 7:3 The principle of the optical lever. As the mirror rotates from M to M', the image on the screen moves from I to I'.

7:2 Refraction at Plane Surfaces

A ray of light travelling through a glass block has its path bent as shown in fig. 7:4. At A light enters the glass, is slowed down, and bent towards the normal. At B the change is reversed and the ray emerges parallel to its original direction but laterally displaced.

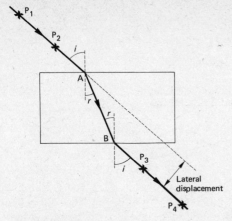

Fig. 7:4 Refraction in a parallel-sided glass block.

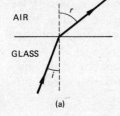

(a)

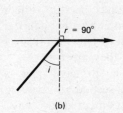

(b)

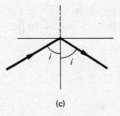

(c)

Fig. 7:6 Total internal reflection at a glass/air boundary.

Fig. 7:5 Deviation produced when a ray passes through a prism.

Measurement of the angles allows us to calculate the refractive index n of the boundary (see section 6:5) since

$$n = \frac{\sin i}{\sin r}$$

A triangular block of glass, called a *prism*, behaves slightly differently since it will produce a net deviation. The path of a typical ray is shown in fig. 7:5. The total deviation D is the sum of the deviations at each refraction.

An interesting and useful effect can arise when light travels from glass to air. Fig. 7:6(a) shows such a situation where

$$r > i$$

As i increases, a situation is reached where r becomes 90° (fig. 7:6(b)). Any further increase in i results in **total internal reflection** within the glass (fig. 7:6(c)) and the boundary acts like a mirror. The angle of incidence at which the change from refraction to reflection occurs is called the **critical angle**. For glass of refractive index 1.5 the critical angle is around 42° and so a 45/90 prism can be used for reflection as illustrated in fig. 7:7. Total internal reflection within prisms is used in binoculars in preference to mirrors.

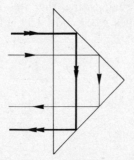

Fig. 7:7 Total internal reflections within a prism. Two reflections turn light through 180° and produce image inversion.

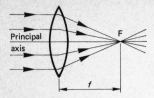

Fig. 7:8 Focussing effect of a biconvex lens. F is the principal focus and *f* is the focal length.

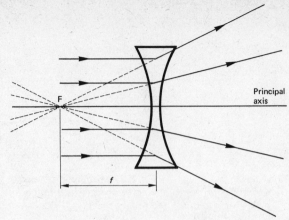

Fig. 7:9 Diverging effect of a biconcave lens showing the virtual principal focus F and focal length *f*.

7:3 Image Formation in Lenses

Two simple lenses are shown in figs. 7:8 and 7:9. The lenses can be described by their shape or by their behaviour. Fig. 7:8 shows a **convex lens** which acts as a *converging* lens. Rays initially parallel to its principal axis undergo two deviations due to refraction at the lens surfaces and are focussed to a point F called the **principal focus**. In fig. 7:9, a **concave lens** is seen to act as a *diverging* lens with rays appearing to diverge from a virtual principal focus F. In each case the distance from lens to principal focus is called the **focal length** *f*.

The value of the focal length for a particular lens depends on the shape of the lens, i.e. the curvature of the surfaces, and also on the refractive index of the glass from which the lens is made.

1 We will now look at **image formation** in converging and diverging lenses. In fig. 7:10 an object O is positioned distance *u* (the object distance) from a **converging lens**. The object distance is greater than the focal length. To locate the image we follow the path of two rays from a point on the object through the lens and see where they meet.

A ray from the top of the object parallel to the principal axis is deviated and passes through the principal focus. A second ray passing through the centre of the lens is undeviated. These rays meet at the image position I. The distance *v* shown in fig. 7:10 is the image distance. If O is an illuminated object a focused image will be seen on a screen placed at I. The image is therefore real. Notice that this image is upside down (or inverted). The two triangles ABD and GED in fig. 7:10 have equal angles and so are similar. Hence

$$\frac{GE}{AB} = \frac{ED}{BD}$$

or, writing the object size as *O* and the image size as *I*,

$$\frac{I}{O} = \frac{v}{u} = m$$

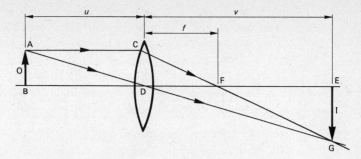

Fig. 7:10 Real image formation in a converging lens. The object distance *u* is greater than the focal length *f*.

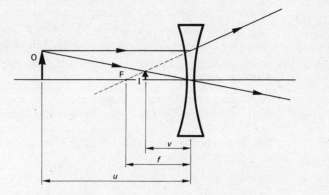

Fig. 7:11 Formation of a virtual image in a diverging lens.

This ratio of sizes is called the **magnification** *m*. If $m > 1$ the image is magnified, while if $m < 1$ the image is diminished.

The triangles CDF and GEF are also similar so that

$$\frac{GE}{CD} = \frac{EF}{DF}$$

Since CD = AB this becomes

$$\frac{I}{O} = \frac{v - f}{f}$$

Substituting for *I/O* from the earlier equation gives

$$\frac{v}{u} = \frac{v - f}{f} \quad \text{or} \quad vf = uv - uf$$

Dividing each term by the product *uvf* gives

$$\frac{1}{u} = \frac{1}{f} - \frac{1}{v} \quad \text{or} \quad \frac{1}{u} + \frac{1}{v} = \frac{1}{f} \qquad (7:1)$$

This equation can then be used to determine the position of an image given the object position.

2 Image formation in a **diverging (biconcave) lens** is shown in fig. 7:11 for an object outside the principal focus. You should check that a similar image is produced if the object is inside the principal focus by drawing your own ray diagram after reading this account.

63

The same two rays as before are used to locate the image. The ray parallel to the axis appears, after deviation, to have come from F, while the ray through the centre is undeviated. The rays emerging from the lens therefore appear to have come from I which is the image position. To a person looking through the lens, O appears to be located at I which is the virtual image. Notice that this image is erect and diminished in size. The image and object distances are again related by the equation

$$\frac{1}{u}+\frac{1}{v}=\frac{1}{f}$$

provided that virtual distances (such as v in fig. 7:11) are taken as negative values and the virtual focal length of the diverging lens is also taken as negative, as illustrated in the following example.

Example 7:1

A biconcave lens forms a virtual image of an object placed 30 cm from it at an image distance of 20 cm. Determine the focal length of the lens.

Solution The image distance $v = $ 20 cm, since it is virtual. Hence substituting in equation (7:1) gives

$$\frac{1}{30}-\frac{1}{20}=\frac{1}{f} \quad \text{i.e.} \quad \frac{1}{f}=\frac{2-3}{60}=-\frac{1}{60}$$

so that $f = -60$ cm

The negative sign indicates that the focus is virtual, i.e. the lens is diverging.

7:4 The Magnifying Glass and Magnifying Power

We have not yet considered what happens when the object in front of a converging lens is inside the principal focus. In this situation the lens acts as a simple **magnifying glass**. The ray diagram is shown in fig. 7:12, and it can be seen that the rays emerging from the lens are diverging. They appear to have come from the virtual image I which is erect and magnified.

Fig. 7:13 shows what happens when the object is located *at* the principal focus. The emerging rays are parallel and so, effectively, the virtual image is at infinity. The infinite image distance makes the magnification expression ($m = v/u$) meaningless.

To deal with this situation we need to think more about image formation. In optical instruments such as the magnifying glass and the microscope, when we talk about magnification we are comparing the size of the object seen with the naked eye with the object as seen in the instrument. In either case, rays enter the eye and are focused by the lens of the eye onto the retina which acts like a screen. Fig. 7:14 shows an object O which forms a retinal image I. If the object moves nearer to the eye (to O′), it will appear larger, that is it will form a larger retinal image I′. At the same time the angle α, which the object subtends at the eye, increases. (This is the angle between rays from the top and bottom of the object.) This angle can be used as a measure of image size.

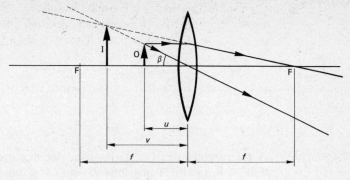

Fig. 7:12 Converging lens as a simple magnifying glass producing a magnified virtual image I. β is the angle subtended by the image at the lens.

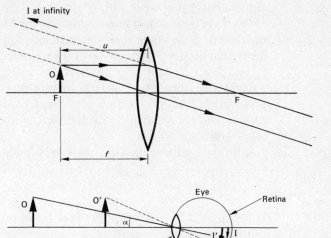

Fig. 7:13 Simple magnifying glass with object at the principal focus. Parallel rays effectively give an image at infinity.

Fig. 7:14 Angle subtended by an object, and retinal images formed in the eye.

In discussing optical instruments, it is usual to define a **magnifying power** for the instrument which is the ratio of the angle subtended by the image produced in the instrument (denoted by β) to the angle subtended by the object when viewed unaided (denoted by α). So, for example, in the magnifying glass shown in fig. 7:12, β is the angle subtended by the image at the lens (near which the eye is located) and

$$\tan\beta = \frac{O}{u} = \frac{I}{v}$$

The human eye can focus objects over a wide range of distances. For normal vision, objects are focused from infinity down to about 25 cm. This latter distance is called the *near point* since it is the closest an object can be to the eye and give a sharp image on the retina. So if the object in fig. 7:12 is viewed unaided at the near point, it will subtend an angle α given by

$$\tan\alpha = \frac{O}{25}$$

In practice the angles are usually small (i.e. the rays are only inclined at small angles to the principal axis) and so the tangent is approximately equal to the angle in radians. The magnifying power M becomes

$$M = \frac{\beta}{\alpha} = \frac{O/u}{O/25} = \frac{25}{u}$$

7:5 The Compound Microscope

A **microscope** can be made from two separated converging lenses, although in practical instruments combinations of lenses replace the individual converging lenses. Fig. 7:15 shows the image formation. A small object O is located just outside the principal focus of the first lens, called the objective, so that a real, inverted and magnified image I_1 is produced. This real image is formed at or just inside the principal focus of the second lens (the eyepiece) which acts as a simple magnifying glass to give a final virtual image I_2. This final image is inverted but this is usually not a problem in the microscope. In normal adjustment, the eyepiece is positioned to form image I_2 at the near point and the overall magnification is then maximum.

In normal adjustment, the angle subtended at the eyepiece by I_2 is

$$\beta = \frac{I_2}{25}$$

If the object was placed at the near point and viewed unaided it would subtend an angle α where

$$\alpha = \frac{O}{25}$$

and so the magnifying power of the microscope becomes

$$M = \frac{\beta}{\alpha} = \frac{I_2/25}{O/25} = \frac{I_2}{O}$$

This ratio could be written instead as

$$M = \frac{I_2}{I_1} \times \frac{I_1}{O}$$

where the first term is the magnification in the eyepiece (m_e) and the second term is the objective magnification (m_o). Hence

$$M = m_o \times m_e$$

Commonly used microscopes have a ×10 eyepiece and often have interchangeable objective lenses, typically with magnifications ×4, ×10, ×40, so that the overall magnifications are ×40, ×100, ×400. For high magnification the eyepiece focal length should be small and, to keep the microscope length reasonable, the objective should also have a short focal length.

Fig. 7:15 The compound microscope. The final image I_2 is formed at the near point. β is the angle subtended by this image at the eyepiece.

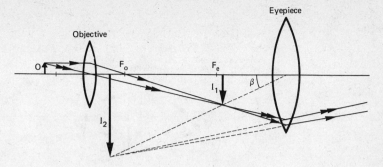

Example 7:2

A microscope with a $\times 10$ eyepiece in normal adjustment produces an overall magnification of $\times 150$. If the lenses are 17.5 cm apart calculate the objective focal length.

Solution   For the eyepiece $m_e = \dfrac{I_2}{I_1} = \dfrac{v}{u}$

i.e. $10 = \dfrac{25}{u}$ so that $u = 2.5$ cm

Since the lenses are 17.5 cm apart, the intermediate image I_1 is formed $(17.5 - 2.5)$ cm from the objective.

From the magnification equation $M = m_o \times m_e$

$$150 = m_o \times 10 \quad \text{so that} \quad m_o = 15$$

For the objective lens

$$m_o = 15 = \frac{v}{u} = \frac{15}{u} \quad \text{giving} \quad u = 1.0 \text{ cm}$$

Finally we can apply the lens equation to the objective to give

$$\frac{1}{f_o} = \frac{1}{1} + \frac{1}{15} = \frac{15 + 1}{15}$$

hence $f_o = \dfrac{15}{16} = 0.94$ cm

Note that, since we know the positions of I_1 and I_2, we could also easily determine the focal length of the eyepiece. You should be able to show that it is 2.78 cm. These short focal lengths give high magnification in a compact instrument.

7:6   The Projector

A schematic diagram of a **slide projector** is given in fig. 7:16. An intense tungsten halogen lamp, which is fan cooled, acts as an extended source to illuminate the slide. The projector lens forms a real image of the slide on a distant screen. For good images, the slide must be brightly and uniformly illuminated. This is achieved by placing it immediately in front of a converging condenser which consists of a pair of plano-convex lenses back to back. The condenser has a short focus so that it can be placed

Fig. 7:16 The optics of a slide projector.

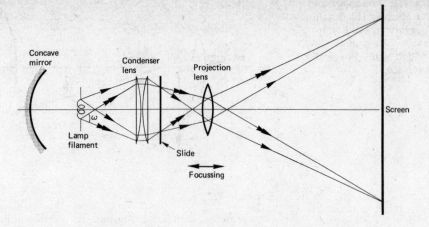

near to the lamp and collect light from a wide angle ω. The illumination is further enhanced by placing a reflecting mirror behind the lamp.

As the figure shows, the condenser forms an image of the source in, or near, the plane of the projector lens, and this lens forms an image of the slide. For the slide to appear the right way up, it must be inserted upside down in the projector. Focusing is achieved by moving the projector lens towards, or away from, the slide which, to give a magnified image, must be just outside its principal focus.

7:7 Dispersion in Prisms

Sunlight is made up of emissions over a continuous range of visible wavelengths. All these emissions travel at the same speed in vacuum. In glass we have seen previously (in section 6:5) that the speed of light is reduced and this leads to refraction at glass/air boundaries. There is, however, an additional effect, called **dispersion,** which occurs in glass.

It is found that the refractive index of glass depends on the wavelength of the light passing through it. The refractive index for violet light is slightly greater than that for red light. Typical figures for crown glass (which is used in lenses) are 1.527 for violet and 1.515 for red. If white light is incident at angle i at a boundary then

$$\frac{\sin i}{\sin r} = n$$

and, because of the variation of n, different colours will have different values of r. The effect when white light passes through a prism is shown in fig. 7:17. Violet light is deviated more than red and dispersion produces a spectrum on the screen. (Spectra will be discussed in more detail in chapter 8.)

Dispersion also occurs in water. A rainbow is produced when sunlight is refracted in small rain drops. Any medium where the speed of a wave is dependent on the wavelength is said to be dispersive.

One consequence of dispersion in glass is that there will be some distortion of the images produced in optical instruments in white light.

Fig. 7:17 Dispersion of white light in a glass prism producing a spectrum.

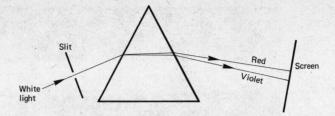

This is called **chromatic aberration** and specially designed lenses known as achromatic doublets are used to reduce the effect.

7:8 Interference of Light

We will now turn to consider the wave properties of light. The general idea of interference for two separate sources has been discussed in section 6:7 so here we will simply describe one example of the interference of light and indicate how it enables us to measure the wavelength.

Interference of light was first demonstrated by Young in 1801, and the principle of his method is indicated in fig. 7:18. The source S should be monochromatic so that either a sodium lamp or a laser would be convenient. The source illuminates a narrow vertical slit and light from

Fig. 7:18 Arrangement of Young's slits, S_1 and S_2, to produce interference pattern on the screen.

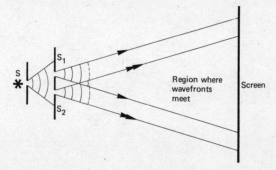

this slit falls onto a blackened screen with two narrow parallel slits S_1 and S_2 scratched on it. These slits, which are typically 1 mm apart, are mounted vertically to be parallel to S. Diffraction occurs at S_1 and S_2 and the emerging light spreads out from each slit. The two slits act like two wave sources and an interference pattern is produced where the wavefronts cross.

A screen is placed 1 m from the source slit and a pattern of parallel, equally spaced bright and dark vertical lines called **interference fringes** is observed. The fringes produced with a sodium source can also be viewed by replacing the screen with an eyepiece which can move across the field of view. (A travelling microscope with the objective removed could be used.) Fig. 7:19 shows the interference fringes formed by a laser source. Fig. 7:19(a) is the central maximum in the single slit diffraction pattern shown in fig. 6:20 and 7:19(b) shows this maximum crossed by the interference fringes produced by waves from the two slits of the Young's arrangement.

Essential Physics

Fig. 7:19 Photograph of Young's interference fringes. Note the equal spacing of the bright and dark fringes. (a) shows the central diffraction maximum which is crossed in (b) by the interference fringes.

Fig. 7:20 Geometry of Young's slits interference.

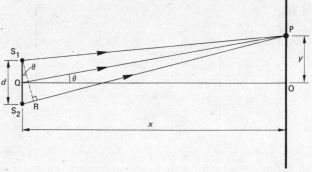

The bright fringes correspond to positions where the waves interfere constructively, while the dark bands occur at positions of destructive interference. At the centre of the pattern, point O in fig. 7:20, the **path difference** is zero and a bright fringe occurs. If P is the position of the adjacent bright fringe, the path difference is

$$S_2P - S_1P = \lambda$$

If the separation of the slits, d, is about 1 mm and the distance from slits to screen, x, is 1 m, the angle θ will be small and we can use small-angle approximations. From the triangles S_1RS_2 and QPO we can write two expressions for θ:

$$\theta = \frac{S_2R}{d} \quad \text{and} \quad \theta = \frac{OP}{x}$$

Since S_1P is approximately equal to RP it follows that S_2R is the path difference λ, while OP is the separation of the bright fringes, y. Equating the expressions for θ and substituting for S_2R and OP gives

$$\frac{\lambda}{d} = \frac{y}{x} \quad \text{or} \quad \lambda = \frac{yd}{x}$$

Hence if the **separation of the fringes** is measured, say by using a micrometer eyepiece, the wavelength can be found. Note that it follows from this equation that, as the slit separation d increases, y will decrease and the pattern closes up, while if x increases and d is kept constant then y will increase. Using a shorter wavelength source will also lead to a decrease in y.

Fig. 7:21 Distinction between a) a continuous wave, e.g. sound from a loudspeaker, and b) photons from a light source. Note (i) the short duration of the pulses, (ii) that individual photons have random phase, (iii) that a source will emit many more photons than is implied here.

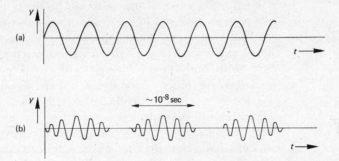

To observe interference of light it is necessary to use a single source, rather than two separate sources. This is because light does not consist of a continuous wave but of short-duration pulses of radiation (see fig. 7:21) called **photons**, each of which results from an individual electron energy change in an atom. Because the photons from separate sources have randomly related phases, the phase relationship between separate sources will continually change and no interference pattern is seen. Two separate light sources are said to be *incoherent*. (We shall return to the production of light in section 8:1.)

Example 7:3

Young's slits 1 mm apart form an interference pattern in mercury green light in a plane 1.2 m from the slits. The fringes are viewed with an eyepiece which moves 6.55 mm in traversing 10 fringes. Find the wavelength of the light.

Solution Since the fringes are equally spaced in a Young's pattern, the separation of adjacent fringes, y, will be

$$\frac{6.55}{10} \, \text{mm} = 6.55 \times 10^{-4} \, \text{m}$$

Substituting for screen distance x, and slit separation d, in the equation

$$\lambda = \frac{yd}{x} \quad \text{gives} \quad \lambda = \frac{6.55 \times 10^{-4} \times 1 \times 10^{-3}}{1.2}$$

with all distances in metres.
Hence $\lambda = 5.46 \times 10^{-7} \, \text{m}$ or $546 \, \text{nm}$

7:9 Diffraction of Light

In the earlier discussion of diffraction in section 6:8 it was indicated that a narrow slit, of size comparable to the wavelength, significantly modified the flow of energy. We will now outline another example of diffraction which also provides a useful way of measuring wavelengths. A **diffraction grating** consists of a large number of closely-spaced narrow slits each of which can be treated as effectively a point (or strictly a line) source. The gratings frequently used in educational laboratories are called *replica gratings* and are produced by ruling closely spaced parallel lines across a metal surface and then taking a plastic casting which is mounted onto a flat glass plate. Light incident on the grating will only pass through in those positions where the original surface remained intact (fig. 7:22).

When a plane wavefront of monochromatic light is incident normally on a grating as shown in fig. 7:23, the slits diffract the light and each slit acts like a separate source producing an emergent wavefront which is curved as shown. Fig. 7:24 shows light diffracted from adjacent slits at an angle θ to the straight-through position. These parallel rays may be focused on the retina of the eye or viewed through a telescope. In either case the waves from individual slits will interfere and, in general, path differences will be such that destructive interference results. However if the path difference between adjacent slits, AC, is equal to a whole number of wavelengths, the waves interfere constructively and a bright image is formed. The path difference AC is given from triangle ACB by

$$AC = AB \sin \theta = d \sin \theta$$

where d is the distance between slit centres and is called the *grating element*. (A typical grating may have 300 lines per mm.)

If θ_1 is the smallest angle at which reinforcement occurs, the path difference is one wavelength and hence

$$\lambda = d \sin \theta_1$$

As θ is increased, further positions of constructive interference may be reached. The next position will correspond to a path difference of 2λ so that

$$2\lambda = d \sin \theta_2$$

and in general $\quad n\lambda = d \sin \theta_n$

This is the grating equation and n, which is an integer, is called the *order*. It follows from the grating equation that if the spacing d is known and the angles for different order images measured, the wavelength of the light can be found.

In most cases the incident light will consist of a number of wavelengths (e.g. either a line or continuous spectrum—see section 8:1). The grating then separates the constituent colours since different wavelengths will reinforce at different angles. This effect could be seen with the arrangement shown in fig. 7:25(a) where a vertical slit is illuminated from a lamp or discharge tube and viewed through a grating held close to the eye. If a white light source is used, the appearance of the image will be as shown in fig. 7:25(b). A white image of the slit is seen at the centre (where all

Fig. 7:22 Diffraction gratings: *a*) section through a metal plate with rulings across it; *b*) replica acts as a series of slits, width *a*, with centres *d* apart. Light only passes through where the original surface is intact.

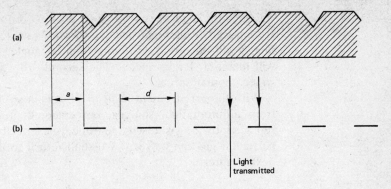

Fig. 7:23 Incident and emergent wavefronts for a plane wave incident normally on a grating

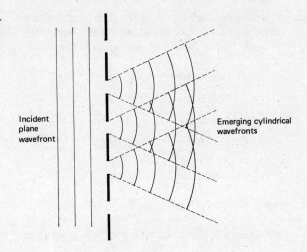

Fig. 7:24 Light emerging at a general angle θ has path difference AC between adjacent slits (The lens is not drawn to scale.)

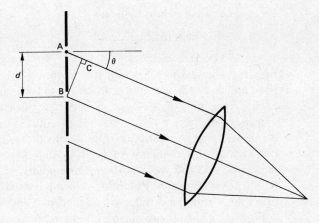

Fig. 7:25 *a*) Simple arrangement to show spectra produced by a grating. *b*) Form of image seen for a white light source.

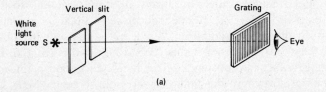

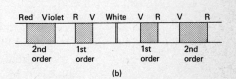

wavelengths reinforce at the same position) but on either side a spectrum is produced. Violet light has the shortest wavelength and so its first-order image will be produced at the smallest angle, while longer wavelengths will image at larger angles. The second-order image is found to have a wider angular spread.

With the arrangement in fig. 7:25 it is possible to compare the spectra from different light sources. For example, if the white light source is replaced by a gas discharge tube, the spectrum characteristic of the particular gas can be seen. We shall return to the topic of spectra in the next chapter.

Example 7:4

A hydrogen discharge tube has strong emissions at wavelengths 656 nm (red) and 486 nm (blue). Calculate the angular separation of the second-order images produced by a diffraction grating with 600 lines/mm.

Solution   The grating spacing here is

$$d = 1/600 \text{ mm} = 1.67 \times 10^{-6} \text{ m}$$

For the 486 nm line, taking $n = 2$ (second order), the grating equation gives

$$2 \times 486 \times 10^{-9} = 1.67 \times 10^{-6} \sin \theta$$

hence $\sin \theta = \dfrac{972 \times 10^{-9}}{1.67 \times 10^{-6}} = 0.583$ i.e. $\theta = 35.69°$

For the 656 nm emission in the second order

$$\sin \theta' = \frac{2 \times 656 \times 10^{-9}}{1.67 \times 10^{-6}} = 0.787 \quad \text{i.e. } \theta' = 51.95°$$

The angular separation is therefore

$$51.95 - 35.69 = 16.26°$$

7:10   Polarization of Light

In a transverse wave motion, the vibrations are restricted to a plane at right angles to the wave direction but can be in any direction in this plane. Light, consisting of a stream of separate pulses, will generally have the vibrations of these pulses randomly orientated in the plane and such light is said to be *unpolarized*. Passage of light through certain crystals however can change the character of light so that the vibrations are all in one direction, and the wave is then *plane polarized*. This effect occurs when light passes through **Polaroid**, which consists of tiny crystals of, for example, quinine iodosulphate aligned in a sheet of nitrocellulose.

Fig. 7:26 shows unpolarized light from a source S passing through Polaroid. The Polaroid transmits vibrations in one direction so the emerging light is plane polarized. If a second piece of Polaroid is placed behind the first, in the same orientation, it will freely transmit the plane polarized light (case (a)). However if this second Polaroid is rotated, the transmitted intensity falls and becomes zero when the rotation is through 90°. In this orientation (where the Polaroids are said to be crossed), the

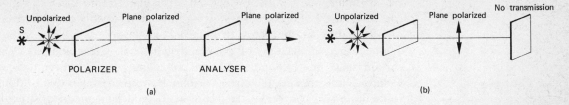

Fig. 7:26 Production and analysis of polarized light. *a)* Analyser transmits vertical vibrations. *b)* Rotation of analyser through 90° prevents transmission of vertical vibrations and no light emerges.

second Polaroid would only transmit vibrations in a horizontal direction. Since the only vibrations reaching it are vertical, no light passes through (case (b)).

This experiment can be easily performed by holding two sheets of Polaroid in front of a light source, rotating one, and observing the resulting change in transmitted intensity. This technique provides a method of detecting polarized light and the second Polaroid acts as an *analyser*. The fact that light shows such behaviour is evidence that it is a transverse wave motion.

Some crystals, such as calcite, show an effect called **double refraction** in which light passing through the crystal is split into two components both of which are plane polarized. A **Nicol prism** is a form of calcite crystal modified so that only one of the components is transmitted.

Nicol prisms act as the polarizer and analyser in a polarimeter which can measure the optical activity shown by materials such as quartz, turpentine and a large number of other organic compounds including sugar solutions. These substances rotate the plane of polarization of polarized light passing through them. In solutions, the angle of rotation is proportional to the concentration which can therefore be determined by measuring the angle.

Fig. 7:27 Schematic diagram of a simple polarimeter

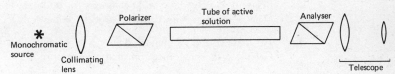

The layout of a simple **polarimeter** is shown in fig. 7:27. A monochromatic source is used (since optical activity depends on wavelength), and the solution is placed between two Nicol prisms which are initially crossed so that no light is transmitted through the analyser. When the solution is introduced, the analyser is rotated through a measured angle to restore darkness, and the concentration can then be determined.

Polarized light can also be produced by reflection. Unpolarized light incident on a glass surface as shown in fig. 7:28 is partially reflected and partially refracted into the glass. When the angle between the reflected and refracted rays is 90°, it is found that the reflected ray is plane polarized with the vibrations in a direction normal to the plane of the diagram. The angle of incidence in this case is called the *Brewster angle*. At a general angle of incidence, the reflected ray, like the refracted ray, is partially plane polarized (so that vibrations in other directions are re-

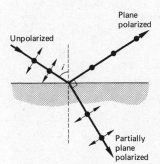

Fig. 7:28 Plane polarization resulting from reflection. Reflected and refracted rays are at right angles and *i* is the Brewster angle.

75

duced rather than completely cut out). The glare from reflecting surfaces can be reduced, therefore, by viewing through a suitably orientated sheet of Polaroid, as is the case with Polaroid sunglasses.

One other application of polarized light is in stress analysis. When certain materials like perspex and polythene placed between crossed Polaroids are stressed, the stress regions show up as fringes due to double refraction. Such materials can be used to model engineering structures which can then be examined to locate the stress points in the structure.

Exercises

7.1   Two plane mirrors, both with their planes vertical, are inclined at 90°. Show, by drawing the path of rays from an object, that three images can be formed in the mirrors.

7.2   Light falling onto a glass surface at an angle of incidence of 50° is partly reflected and partly refracted. The glass has refractive index 1.53. Calculate the angles of reflection and refraction. Calculate also the angle between the reflected and refracted rays.

7.3   If the critical angle at a diamond/air boundary is 24.5° determine the refractive index of diamond. Hence determine the speed of light in diamond. (Velocity of light in air $= 3.0 \times 10^8$ m/s.)

7.4 a) An object is placed 60 cm from a converging lens of focal length 20 cm. Find the position, nature and size of the image formed. b) Explain the corresponding situation of the image if the object was placed 10 cm from the lens.

7.5   A beam of light is converging towards a point P on the axis of a biconcave lens which lies 20 cm beyond the lens. The lens focal length is 20 cm. Use a ray diagram to determine what happens to the beam after passage through the lens. (The point P acts as a virtual object for the lens.)

7.6   A small object is placed 8 cm from a diverging lens of focal length 24 cm. Calculate the magnification of the image produced.

7.7   Explain which of the following appears to be larger when viewed unaided (i.e. which produces the larger image on the retina of the eye):
a) the moon, which subtends an angle of 0.5°
b) a 15 m high building at a distance of 800 m.

7.8   The magnifying power of a magnifying glass is given in section 7:4 as $M = 25/u$. Show that if the virtual image I in fig. 7:12 is formed at the near point (25 cm) the magnifying power can be written in terms of the focal length f of the lens as

$$M = 1 + \frac{25}{f}$$

(Hint: apply the lens equation.)

7.9   A compound microscope has an objective of focal length 1 cm and an eyepiece of focal length 5 cm. The object is located 1.05 cm from the objective and the final image is 25 cm from the eyepiece. Find the separation of the lenses and the magnifying power of the microscope.

7.10 A 35 mm slide is to be projected to fill a screen 1.05 m wide. If the maximum distance from the projector lens to the screen is 5 m calculate what the focal length of the projector lens would need to be.

7.11 White light is incident on the face of a prism at an angle of incidence of 45°. The refractive index of the glass varies from 1.639 for violet light to 1.612 for red light. Calculate the angle between the extreme refracted rays at the first face.

7.12 Young's fringes are observed in blue hydrogen light (wavelength 486 nm) which illuminates two slits 1 mm apart. Determine how far the viewing plane is from the slits if the fringe separation is 0.5 mm.

7.13 A diffraction grating with 600 lines per mm forms diffracted images with light of wavelength 486 nm. Determine the highest-order diffracted image observable for this wavelength.

7.14 Explain which wavelength in the third-order spectrum produced by a diffraction grating would coincide in angular position with the second-order image for red light of wavelength 656 nm.

7.15 Calculate the Brewster angle for dense flint glass of refractive index 1.65.

8 Spectra and the Electromagnetic Spectrum

Sources of light differ according to whether they emit over all visible wavelengths, in which case they are said to produce a **continuous spectrum**, or whether certain wavelengths only are emitted, giving a **line spectrum**. Sunlight and the emission from filament lamps are examples of a continuous spectrum, while discharge tubes and sodium lamps produce line spectra.

In the first part of this chapter we will be looking at different types of spectra and how they are produced. The second part of the chapter is concerned with the electromagnetic spectrum and the distinctive properties of different regions within it.

8:1 Emission Spectra

The line and continuous spectra mentioned above can be seen by examining the output from a source using, for example, a diffraction grating. Such spectra are called **emission spectra**.

The emission of light results from energy changes within **atoms**. The electrons in atoms can only have certain energies, and any changes in energy must be between these allowed energy values. In a collision an electron may gain, or absorb, energy. This excitation leaves the electron in a higher energy state which is unstable. The electron therefore moves back to a lower energy state and releases its excess energy as a pulse of light (or other electromagnetic radiation). The wavelength λ, or frequency f, of the emission is related to the energy change. If an electron moves from energy E_B to a lower energy value E_A then

$$E_B - E_A = hf = \frac{hc}{\lambda}$$

where h is a constant called Planck's constant (6.63×10^{-34} J s) and c is the speed of light. These processes will be discussed more fully in chapter 19, and are illustrated in fig. 19:7 (page 210).

It follows from the equation above that, if E_A and E_B can only take certain values, there will only be certain wavelengths emitted. The values of E_A and E_B depend on the particular element involved and so the emitted wavelengths will be characteristic of a given element. Each element has its own spectral fingerprint. Fig. 8:1 shows the major spectral lines for a number of elements. Line spectra of this sort are produced from isolated atoms in gases or vapours. In a sodium source, the emission at 589 nm (yellow light) is so much more intense than the other emissions that it may be regarded as a single wavelength (or monochromatic) source.

Fig. 8:1 Examples of the main emission lines in atomic spectra. Wavelengths are quoted in nanometres. The dashed lines show the extent of the visible region (400–700 nm).

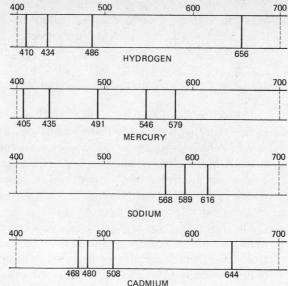

HYDROGEN

MERCURY

SODIUM

CADMIUM

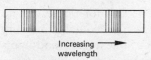

Increasing ⟶ wavelength

Fig. 8:2 The form of the bands in a band spectrum. The lines are closer together and the emission more intense at one end of the bands.

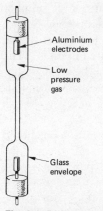

Aluminium electrodes

Low pressure gas

Glass envelope

Fig. 8:3 Discharge tube used for producing emission spectra. A high potential difference is connected across the electrodes.

The spectra produced by **molecules** are somewhat more complex. The close proximity of atoms within the molecule slightly modifies the allowed energies, and changes between levels produce a band corresponding to a narrow range of wavelengths rather than a line. The bands consist of a number of closely spaced lines as shown in fig. 8:2. **Band spectra** are produced by molecular gases such as oxygen (O_2) and nitrogen (N_2).

In solids and liquids the situation is further complicated because of the influence of adjacent atoms, and the bands in fig. 8:2 are spread into a continuous emission over a wide range of wavelengths. In the domestic light bulb, for example, a filament of tungsten is heated to a high temperature so that it gives off emissions over the whole of the visible range. Its spectrum is then continuous.

8:2 Production and Use of Spectra

As we have seen, a continuous emission spectrum is produced by a filament lamp, and also by the sun. Line and band spectra can conveniently be produced in a **discharge tube** (fig. 8:3). The glass tube contains gas (such as hydrogen or helium) at very low pressure (in the range 0.5–10 mm of mercury). A high potential difference (typically about 3000 V) is applied across the tube producing electrical conduction and consequent excitation of the gas. A visible glow is produced which will be most intense in the narrow section of the tube.

A neon discharge tube emits a characteristic red colour from its particular line spectrum which is utilised in the neon signs for advertising. Other colours can be produced by adding mercury to the inert gas and by using coloured glass.

Characteristic line spectra are also produced in vapour lamps using, for example, sodium, mercury and cadmium. In these lamps, heat generated when the lamp is first switched on vaporizes the metal. The subsequent passage of current through the vapour produces a line emission. The orange/yellow street lamps used to illuminate main roads and motorways are similar to the sodium lamps used in laboratories and also emit a line spectrum.

The other familiar light source which is used both for domestic and commercial lighting is the **fluorescent tube**. This tube contains mercury vapour which produces a line emission in the visible region and emits strongly in the ultra-violet. This UV radiation is absorbed by the chemical coating on the inside of the tube, and the energy is re-emitted at visible wavelengths. (This process is called fluorescence.) With a suitable mixture of chemicals a white light emission is produced.

The **laser** is another source of monochromatic radiation with wide industrial and research applications. It utilises the effect of stimulated emission in which simultaneous emission of energy from a large number of atoms occurs with individual photons in phase with one another. Such a source is said to be *coherent*. This is quite different from a conventional light source where individual photons are randomly related in phase. The device produces a narrow (i.e. highly collimated) intense beam of radiation at a single wavelength. A He/Ne gas laser gives a continuous emission with a power output typically of 1 mW or more at a wavelength of 633 nm.

The intensity of the laser emission necessitates care in use. You should never look along the laser beam or view reflections off a mirror or shiny surface since the beam may damage the retina. With powerful lasers, special spectacles should be worn.

Spectra can be produced and examined either by making use of the dispersion effect in glass (illustrated in fig. 7:17) or by means of a diffraction grating. In a **spectrograph**, light passes through a prism and is then focused so that different wavelengths produce an image at different points on a photographic plate. The photograph is called a spectrogram. Another instrument, called a direct vision spectroscope, uses a combination of prisms so that dispersion is produced without any deviation. The spectroscope can then be pointed directly at the source and the spectrum observed. The instrument usually has a scale incorporated so that wavelengths can be read off directly.

Each element has its own characteristic spectrum so that the appearance of these wavelengths indicates the presence of the element. Furthermore the intensity of the lines gives a measure of the quantity of the element present. A knowledge of the characteristic wavelengths of an element also gives information about the energy levels within the emitting atoms.

8:3 Absorption Spectra

In an emission spectrum, light is viewed directly from the source. A different type of spectrum, called an **absorption spectrum**, is produced

Fig. 8:4 Experimental arrangement for viewing an absorption spectrum.

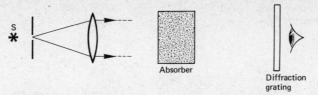

Absorber

Diffraction grating

when light passes through an absorbing medium placed between the source and the observer. In fig. 8:4 a white light source S illuminates a slit at the principal focus of a convex lens. The lens produces a collimated beam (i.e. parallel rays) which passes through an absorber before being viewed in a grating. The spectrum consists of the continuous emission of the source crossed by dark lines or bands at certain wavelengths which are characteristic of the absorber. The absorber could be iodine or sodium vapour.

When the light passes through the vapour, certain wavelengths are absorbed. These are the wavelengths which correspond to the energies needed to excite electrons in the vapour atoms to higher levels. The atoms then re-radiate the same wavelengths in random directions so that the intensity arriving at the grating is much reduced and the spectrum appears dark at these wavelengths. Elements absorb the same wavelengths that they characteristically emit. We could regard the absorption spectrum as the negative of a photograph for which the emission spectrum is the positive or print.

The continuous spectrum of sunlight which reaches the earth is crossed by a series of dark lines, called Fraunhofer lines, resulting from absorption by elements such as hydrogen, helium, sodium, calcium and iron in the solar atmosphere.

8:4 The Electromagnetic Spectrum

The **electromagnetic spectrum** is the term used for a family of waves which all have the same fundamental character although they show widely diverse properties. We will look first at the common features before considering the distinctive behaviour of the various members of the family.

When an electric charge is accelerated, disturbances are set up in its electric and magnetic fields and these disturbances travel through space with a characteristic speed. This combination of electric and magnetic fields in space constitutes an **electromagnetic wave**. The direction of the fields relative to the wave direction is shown in fig. 8:5. The electric and magnetic fields which vary sinusoidally are always perpendicular to the wave direction (so that the wave motion is transverse).

Electric and magnetic fields can be set up in a vacuum so that electromagnetic waves, unlike mechanical waves which rely on a medium, can travel through a vacuum. The speed of all types of electromagnetic waves is the same in a vacuum and is given by

$$c = \frac{1}{\sqrt{[\mu_0 \epsilon_0]}}$$

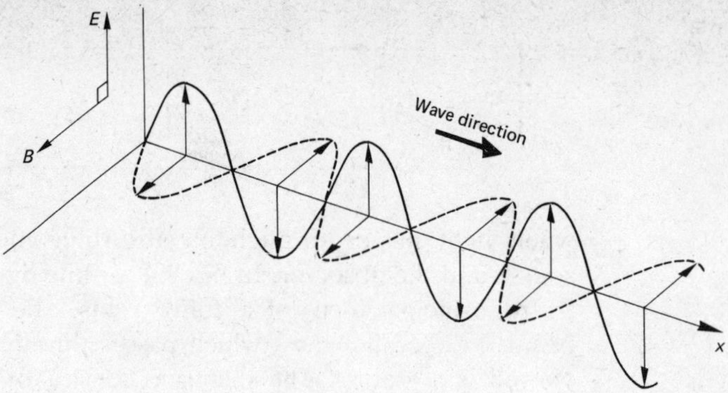

Fig. 8:5 Electric and magnetic fields (E and B respectively) in an electromagnetic wave travelling along the x-direction.

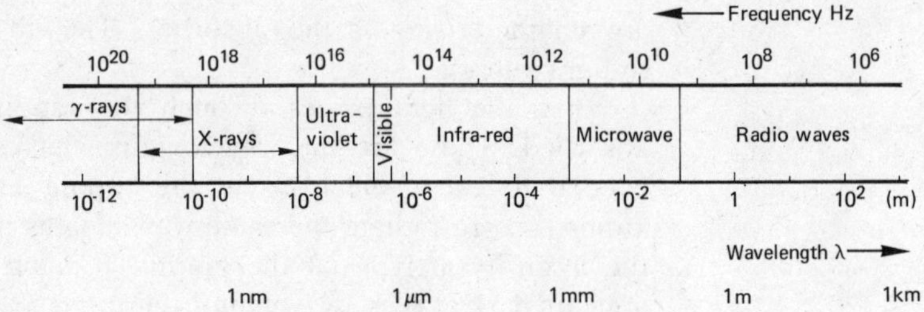

Fig. 8:6 The electromagnetic spectrum, showing the main divisions.

where μ_0 is the *permeability* of free space, or vacuum ($4\pi \times 10^{-7}$ H/m), and ϵ_0 is the *permittivity* of free space (8.854×10^{-12} F/m). These constants will be met again in sections 13:1 and 15:7. With the values given you can show that the speed of the waves in vacuum is 3×10^8 m/s.

The major divisions of the electromagnetic spectrum are shown in fig. 8:6 with an indication of the wavelengths and frequencies (which are related by $c = f\lambda$). The divisions are only approximate, so that, for example, the difference between γ-rays and X-rays with wavelength 10^{-11} m is simply in the way the radiation is produced, not in their properties.

8:5 Gamma-radiation and X-radiation

The general character, properties and applications of gamma-rays (or γ-rays) and X-rays are similar, but the methods of production are quite different. Gamma-rays are one of the emissions from radioactive materials. They are emitted as a result of energy changes within the atomic nucleus. X-rays are produced in a special type of vacuum tube.

Both gamma-radiation and X-radiation are detected in similar ways. Some detectors depend on ionization which the radiations produce in a gas, while the scintillation counter detects the fluorescence caused by the radiations. Detectors are discussed in detail in chapter 22.

Gamma-rays and X-rays are both widely used in medicine. Gamma-radiation is used in radiotherapy, while X-rays are used mainly for diagnostic work in radiography. Longer-wavelength (and lower-energy) X-rays, often called soft X-rays, are used in dental photography. More details of gamma-ray applications will be given in section 21:5.

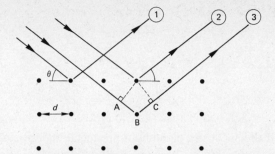

Fig. 8:7 Reflection of X-rays from atomic centres in a crystal.

The typical X-ray wavelength, around 10^{-10} m, is similar to the spacing between atoms in a solid. Since diffraction occurs when the slit size is comparable to the wavelength, we might expect X-rays to be diffracted by crystals. This forms the basis of a widely used technique for studying crystal structures.

The atoms in a crystal are arranged in a regular array and act as a three-dimensional grating. Fig. 8:7 shows the positions of the atomic centres in three layers of atoms in a crystal. X-rays are incident on the top plane at an angle θ to the layer. X-radiation is scattered from the atomic centres which act like weak sources in a similar way to the slits in an optical grating. Radiation travels from these centres in all directions. At the reflection angle, radiations 1 and 2 in fig. 8:7 will be in phase and constructively interfere. There will however be a phase difference between this radiation and the radiation (such as 3) reflected off a lower plane of atoms. The path difference between 2 and 3 is AB + BC and if the atomic planes are distance d apart then

$$AB = BC = d \sin \theta$$

and the path difference is $2d \sin \theta$. At certain values of θ the radiation from all planes will be in phase, constructive interference occurs, and an intense reflection results. The condition for this is that

$$2d \sin \theta = n\lambda$$

where n is an integer called the order of reflection. This relationship is called the Bragg equation. From measurements of θ using monochromatic X-rays (i.e. of a single wavelength) information is obtained about the atomic spacing in crystals.

Example 8:1

X-rays of wavelength 1.5×10^{-10} m are incident on a crystal and produce a first-order reflection at 15°. Determine the crystal spacing and the angle at which the second-order reflection occurs.

Solution Both parts of the question can be answered by applying the Bragg equation. For the first-order reflection substituting the known values gives

$$1 \times 1.5 \times 10^{-10} = 2d \sin 15°$$

$$\therefore \quad d = \frac{1.5 \times 10^{-10}}{2 \times 0.259} = 2.90 \times 10^{-10} \text{ m}$$

For the second-order reflection off the same planes

$$2 \times 1.5 \times 10^{-10} = 2 \times 2.90 \times 10^{-10} \sin\theta$$

so that $\quad \sin\theta = \dfrac{2 \times 1.5}{2 \times 2.9} \quad$ giving $\quad \theta = 31.1°$

8:6 Ultra-violet (UV) and Infra-red (IR) Radiation

If the spectrum of sunlight produced by a prism is photographed, the film is found to be fogged either side of the visible region. This fogging indicates the presence of radiation not detectable by the human eye. Beyond the short wavelength end of the visible region lies the **ultra-violet radiation**, which is also produced by electron energy changes within *atoms*. In producing UV however the energy jumps are larger.

The emission of **infra-red radiation**, with wavelengths longer than visible light, results from energy changes in *molecules*, with either the vibrational or rotational kinetic energy decreasing.

Neither UV nor IR spectra can be studied using components made of glass because glass absorbs wavelengths less than about 350 nm and becomes opaque in the near IR. The opacity of glass accounts for the greenhouse effect. Most of the solar energy at visible or near IR wavelengths can penetrate into the greenhouse and is absorbed by the soil and plants. These bodies re-emit energy but at much longer wavelengths which the glass will not transmit. The solar energy is therefore trapped and the temperature rises.

1 UV-radiation can be detected photographically, by the use of fluorescent screens or by photoelectric devices such as the photomultiplier tube, fitted with a quartz window. The operation of this device will be described in section 22:3. Exposure to longer wavelength UV can produce tanning of light-coloured skins due to increased amounts of melanin (a dark-coloured pigment) which forms a protective layer over skin cells to filter out the UV. The earth is protected from the more harmful shorter wavelength UV (less than 290 nm) which is absorbed by ozone in the upper atmosphere.

2 The IR region is most commonly associated with heat. Bodies, even at ordinary temperatures, radiate heat, or infra-red radiation. At higher temperatures, some of the radiant energy is emitted at the red end of the visible region. At any particular temperature, the energy spectrum (that is the energy being emitted at different wavelengths) has the form shown in fig. 8:8. An increase in temperature leads to an increased emission at all wavelengths. This curve and its significance in the development of quantum theory will be considered in section 18:2.

IR-radiation can be detected by its heating effect. One type of electrical detector depends on change of resistance as IR radiation is absorbed. The radiation can also be detected photographically or by means of a photo-transistor. IR lamps are used in treating muscular complaints, and the

Fig. 8:8 Energy spectrum from a hot body at two different temperatures, where T_2 is the higher.

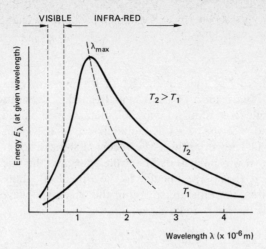

study of the IR emissions from the body can be used for medical diagnosis. Finally, infra-red spectroscopy (looking at molecular absorption spectra) is a standard technique for chemical analysis.

8:7 Microwaves

Microwaves, with typical frequency 10^{10} Hz and wavelength 3 cm, are produced in special electronic devices such as the solid state Gunn diode. This wavelength makes microwaves convenient for demonstrating basic wave properties. For such experiments, a 10.7 GHz (2.8 cm) generator fitted with a horn is used. The receiver consists of a diode detector, either in the form of a small non-directional probe or fitted with a horn, connected to a microammeter.

Fig. 8:9 shows the set-up for two-slit interference. As the probe detector moves along a line parallel to the slits, the strength of the signal rises and falls in a regular way. If the double slit is replaced by a single slit of width reducible from say 30 cm to 3 cm, it is possible to study how the diffraction pattern changes as the width decreases.

Fig. 8:9 Arrangement for showing two-slit interference with a microwave source.

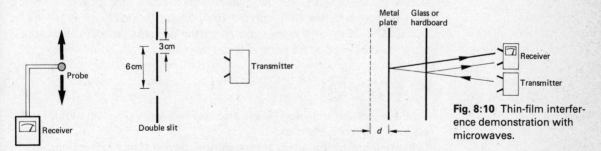

Fig. 8:10 Thin-film interference demonstration with microwaves.

Thin-film interference can be demonstrated by the arrangement in fig. 8:10. The glass or hardboard sheet acts as a partial reflector so that two signals, one from the glass and one from the metal plate, reach the receiver. As the metal plate slowly moves away, the two reflected signals go in and out of phase. If the receiver output goes through n minima while the plate moves distance d, the total change in the path difference is $2d$ and

$$2d = n\lambda$$

where λ is the wavelength of the microwaves.

Polarization of the waves from the transmitter may be shown using a grille consisting of parallel metal wires placed between the transmitter and receiver to act as the analyser. As the grille is rotated as shown in fig. 8:11 the strength of the received signal varies, showing that the original wave was polarised. The grille transmits the microwaves when the electric field E is perpendicular to the wires but, when the field is parallel to the wires, currents are set up in the wires, the wave energy is dissipated, and the receiver signal is minimum.

Fig. 8:11 Wire grille used to show polarisation of microwaves. In case (a) the wave energy is absorbed in producing current flow in the wires.

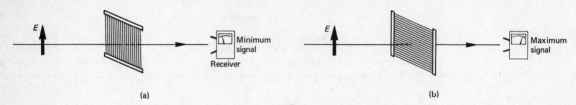

(a)   (b)

Microwaves find application for example in radar and in cookers. In radar, pulses of microwave radiation are transmitted and detected by parabolic dish aerials. The detector signal is usually a weak reflection off a distant moving object. In a microwave oven, food is exposed to waves at a frequency around 2.45 GHz. Microwave energy at this frequency is absorbed by water molecules in the food which therefore heats up. Since the radiation can penetrate several centimetres into the food, it quickly cooks or warms right through instead of cooking slowly from the outside as in a conventional oven.

8:8 Radio Waves

The **radio wave** region, familiar because if its use in communications, covers the range from TV signals to long-wave radio signals. Fig. 8:12 shows the main parts of the range. These waves are generated when high-frequency alternating current from an oscillator circuit is fed to an aerial. The oscillatory circuit incorporating an inductance L and capacitance C oscillates at its natural frequency

$$f = \frac{1}{2\pi\sqrt{[LC]}}$$

Radio waves are detected by electric currents induced in an aerial by the electric field of the waves.

In communication applications, information is sent by superimposing a signal onto the radio wave which acts as the carrier signal. This process is called **modulation** and may be achieved either by varying the amplitude of the carrier (amplitude modulation, AM) or modifying the frequency of the carrier (frequency modulation, FM). The receiver then extracts the information from the carrier to produce an audio output.

Fig. 8:12 The main divisions of the radio spectrum. (v.h.f. is very high frequency, and u.h.f. is ultra high frequency).

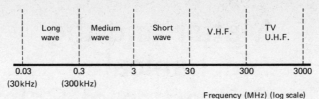

Frequency (MHz) (log scale)

Exercises

Where needed, the speed of electromagnetic waves in vacuum is 3.0×10^8 m/s.
Planck's constant $h = 6.63 \times 10^{-34}$ J s

8.1 Explain the energy change that occurs within an atom which emits a pulse of light of wavelength 486 nm.

8.2 Explain the ways in which the spectra from the following three sources differ:
a) discharge tube containing oxygen
b) car headlamp bulb
c) cadmium lamp.

8.3 The energy required to ionize an oxygen molecule is 1.94×10^{-18} J. Find the longest wavelength of electromagnetic radiation which could ionize oxygen. State which part of the spectrum this is in.

8.4 List three properties common to all forms of electromagnetic waves and three properties which enable different forms to be distinguished.

8.5 Arrange the following electromagnetic emissions in order of increasing frequency:
a) radiation from a hot furnace
b) 30 mm radar pulse
c) carrier wave of a TV signal
d) emission from a cobalt 60 nucleus
e) radiation emitted from an a.c. mains transformer.

8.6 Explain why longer wavelength X-rays (soft X-rays) are less penetrating than those of shorter wavelength.

8.7 X-rays incident on a crystal of lithium fluoride (LiF) where the spacing of the atomic planes is 4.03×10^{-10} m produce a second-order reflection at an angle of 50° to the planes. Calculate the wavelength of the X-rays.

8.8 State which of the following statements are true:
a) Green light has longer wavelength than yellow light.
b) X-rays cannot be transmitted through crystals.
c) Radio waves cannot be polarized.
d) Presence of Fraunhofer lines is evidence for certain elements in the interior of the sun.
e) X-radiation is more penetrating than UV-radiation.
f) X-rays travelling through air produce ionization.

8.9 In the experimental arrangement shown in fig. 8:10, calculate how far the metal plate would move if the detector goes through 16 minima and the transmitter emits 3 cm waves.

8.10 A grille of thin parallel wires 6 cm apart act as a diffraction grating for 3 cm microwaves incident normally on it. Determine the angle of the first-order diffraction.

9 The Nature of Matter and the Mechanical Properties of Solids

9:1 Atomic Bonding

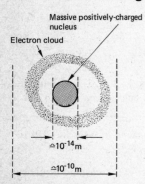

Fig. 9:1 Pictorial representation of a neutral atom.

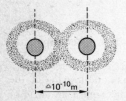

Fig. 9:2 Two neighbouring atoms showing overlap of electron clouds leading to possibility of bonding.

All matter is composed of atoms. In this introductory section we shall show in outline how atoms can be bonded together to form molecules and solids.

The conventional picture of an atom is represented by a massive positively-charged **nucleus**, consisting of *protons* and *neutrons* held together by very strong nuclear forces. The nucleus is itself surrounded by a cloud of **electrons**, which are conventionally considered to be light negatively-charged particles. The electrons in the cloud are in rapid motion around the nucleus and their distances from it will not remain constant (see fig. 9:1). In an electrically-neutral atom, there will be as many protons (units of positive charge) in the nucleus as there are electrons (units of negative charge) in the surrounding cloud. This number is the **atomic number** Z, which determines the chemical nature of the atom and its position in the periodic table. For example, the "sodium-ness" of sodium depends on the fact that $Z = 11$ for all sodium atoms. (Atomic structure will be considered in detail in Chapter 19.)

The bonding which occurs between neighbouring atoms comes about as a result of the inter-meshing of the outermost electrons associated with the two atoms. The distribution of electrons around the nucleus has already been described loosely as a "cloud"; it is the overlap of these "clouds" that produces the bonding (see figs. 9:1 and 9:2). The forces which are involved in this process are thus electric and magnetic in origin. In this way, individual atoms are bonded together to form molecules, and, on the larger scale, atoms and/or molecules are bonded together to form bulk solids.

9:2 Crystalline Solids

All of the solids that will be considered here are crystalline in nature. This means that the atoms or molecules which form the basic building blocks of the solid are arranged in a precise but simple geometrical relationship to each other. We may imagine a **crystal lattice** as a regular arrangement of points in space, on which the atoms or molecules are located (see figs. 9:3(a) and (b)). By joining up neighbouring points with straight lines, it is possible to form **unit cells**. In a three-dimensional lattice, various families of parallel planes can be drawn through the points; these planes will form the faces of a three-dimensional unit cell (fig. 9:3(c)). The unit cell is a

Fig. 9:3 (a) A two-dimensional lattice, i.e. an ordered array of points in a plane giving a unit cell in the shape of a parallelogram of sides *a* and *b*.

Fig. 9:3 (b) A three-dimensional lattice. The points marked as • lie in one horizontal plane and those marked × lie in a parallel plane below it.

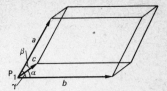

Fig. 9:3 (c) Unit cell for a three-dimensional lattice. The sides are of length *a, b* and *c* with angles α, β and γ between them.

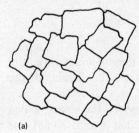

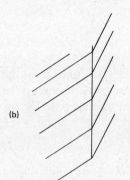

Fig. 9:4 (a) Idea of grain structure. (b) Representation of atom-rich planes at the boundary separating two neighbouring grains.

useful idea because we can think of these as being stacked together, rather in the manner of nursery bricks, to form the crystalline sample. Unit cells take on one of a limited number of simple geometrical shapes, e.g. cubic, hexagonal or rhombohedral. The term *long-range order* is used to denote that the lattice arrangement is maintained over a large number of inter-atomic distances.

It has been assumed so far that the crystal structure is perfect; much of the body of elegant theory of the solid state is based on this assumption. However, in real crystals this is not usually the case and various forms of **imperfection** can occur due to the presence of impurities, vacancies, and atoms being located in the wrong place.

Most metal samples as commonly encountered are *polycrystalline*; here there is a "grainy" structure. This grainy structure can be observed in the freshly fractured surface of a brittle material such as cast-iron, but a more detailed observation can only be made using a microscope with special illumination. Within one grain, order is maintained over, perhaps, several hundred atomic spacings; then in a neighbouring grain, a similar order exists, but with the planes containing the atoms in a slightly different orientation as illustrated in fig. 9:4.

Materials that do *not* exhibit a crystalline structure are said to be *amorphous* (derived from Greek words meaning "without shape" or "without form"). Typical examples of amorphous materials are sulphur and glass; glass, however, tends to take on a crystalline form as it ages over a period of many years.

9:3 Deformations in Solids

It is well known from dynamics that, if a force acts on a body, an acceleration will be produced *provided that the body is free to move.* However if the body is constrained from moving—say by being rigidly clamped at some point—the effect of a force acting on it will be to *deform* the body in some way.

The simplest example is a long straight wire which hangs vertically from a rigid support at the top and which carries a load at its lower end as in

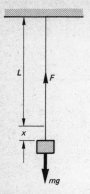

Fig. 9:5 Load *m* in equilibrium supported by a vertical wire of unstretched length *L*. The extension produced is *x*.

fig. 9:5. Gravitational attraction on the load pulls it down, causing the wire to extend slightly, and sets up a tension in the wire which is equal and opposite to the weight when the load hangs in equilibrium. Typically, for a steel wire of length 2.0 m and diameter 0.5 mm carrying a load of 5.0 kg, an extension of some 2 to 3 mm might be expected. In such a case we may think of the external force (the weight due to the applied load) doing work against the internal electric and magnetic forces which are producing the bonding between the atoms in the material of the wire. The deformation is a simple extension, an increase in the length of the wire. In practice we should also find that the increase in length of the wire is accompanied by a slight decrease in its diameter, but this is an effect that we can reasonably ignore at this point.

It is convenient to define a quantity called the **strain**, which is the relative (or fractional) deformation. In the case of the long vertical wire:

$$\text{Strain} = \frac{\text{Increase in length}}{\text{Original length}} \qquad (9:1)$$

Clearly the strain, being a ratio of two lengths, is a dimensionless quantity and, from the typical values quoted above, we can see that strains of the order of 1 part in 1000 (2 to 3 mm in 2 m) might be expected.

It is usual to consider the deforming force in terms of the **stress**, which is the deforming force divided by the area over which it acts. Again, for the long vertical wire:

$$\text{Stress} = \frac{\text{Tension due to the load}}{\text{Cross-section area of the wire}} \qquad (9:2)$$

The units for stress are newtons per metre squared (N/m^2) and, from the figures quoted in the example, it is clear that stresses in excess of $10^8 \, N/m^2$ can be supported by a steel wire. It is left as an exercise for the student to verify this statement.

Example 9:1

A wire of unstrained length 2.5 m and cross-sectional area $5 \times 10^{-6} \, m^2$ shows a strain of 10^{-4} when loaded by a mass of 4 kg. Determine the stress in the wire and the extension when loaded.

Solution   The deforming force on the wire is equal to the weight of the load, i.e. the force is

$$4 \times 9.8 = 39.2 \, N$$

From equation (9:2) the stress is given by

$$\frac{39.2}{5 \times 10^{-6}} = 7.84 \times 10^6 \, N/m^2$$

The extension may be found from the strain by substituting into equation (9:1) which gives

$$10^{-4} = \frac{x}{2.5} \quad \text{i.e.} \quad x = 2.5 \times 10^{-4} \, m = 0.25 \, mm$$

9:4 Elastic Deformations

The use of the word *elastic* in physics and materials science is very different from that in everyday non-scientific life. Hard steel is a good example of an elastic material in the sense in which physicists and materials scientists use the term; this seems a far cry from the rubber cord, usually called elastic, used in the clothing industry. The non-scientist associates the term "elastic" with rubber simply because it stretches well; this does not provide a good enough definition for our purposes. The more precise definition which we require depends on the relationship between stress and strain, and upon the added condition that the sample returns to its unstrained condition when the applied stress is removed. In terms of the stresses that rubber cord is subjected to in everyday life, it is this second condition that fails to be met.

The stress/strain relationship is expressed in terms of **Hooke's law** which states that for small deformations:

$$\frac{\text{Stress}}{\text{Strain}} = \text{constant} \qquad (9:3)$$

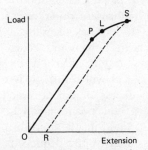

Fig. 9:6 A typical load/extension curve. P denotes the limit of proportionality and L the elastic limit. SR shows the return behaviour when strained beyond the elastic limit giving a permanent set OR.

When a sample is progressively stressed, the strain will increase from zero to a value which, as we have seen, might typically be somewhat greater than 10^{-3}. The change in the strain is proportional to the increasing stress and, provided that the sample returns to its unstrained condition when the stress is completely removed, we describe the deformation process as **elastic**. In a subsequent section we shall see how the stress/strain relationship may be investigated for the case of a long, straight wire but, from what has already been said, it should be clear that for small deformations the stress/strain graph will be a straight line. The point up to which Hooke's law applies is known as the *limit of proportionality*. In some materials, elastic behaviour persists slightly beyond this point, in that when the stress is removed the sample will still return to its unstrained condition. The point at which elastic behaviour breaks down completely is known as the **elastic limit**. Once a sample has been strained beyond the elastic limit, it will not return to the unstrained condition and a *permanent set* is found when the stress is removed (fig. 9:6).

The constant obtained from the Hooke's law stress/strain ratio is known as the **modulus of elasticity**. Clearly, since the strain is dimensionless, the units for the modulus of elasticity will be the same as those for stress, i.e. N/m^2. Three moduli of elasticity are recognized according to the different types of elastic deformation that may be encountered. We shall be concerned with only one of these, *Young's modulus*, which relates to *tensile stresses* and *longitudinal strains* as in the case of a stretched wire under tension. Thus we may define

$$\text{Young's modulus } E = \frac{\text{Tensile (or compressional) stress}}{\text{Longitudinal strain}} \qquad (9:4)$$

91

9.5 A Long Wire Stressed to its Breaking Point

We now consider what happens when the wire as illustrated in fig. 9:5 is **stressed beyond the elastic limit**. Experiment shows that, once elastic behaviour has broken down, further increases in the stress will lead, in many cases, to a relatively large increase in the strain. Materials, such as copper and lead, which exhibit this kind of behaviour are described as *ductile*. A typical load/extension graph for a ductile material is shown in fig. 9:7. The principal features of interest in this graph are

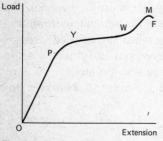

Fig. 9:7 Typical load/extension curve for a ductile material.

the **yield point** Y
the region of **plastic flow** YW
the region of **work hardening** WM
the point of **maximum stress** M
which is rapidly followed by **fracture** at F

Materials like hard steel, however, exhibit elastic behaviour almost to the point of fracture, as shown in fig. 9:8.

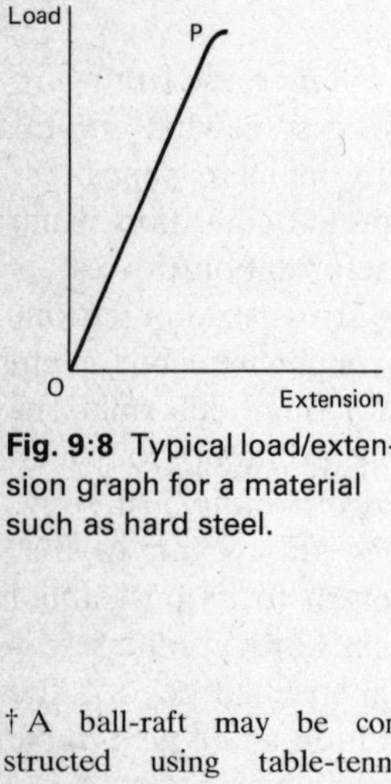

A physical explanation of the various regions of the graph (fig. 9:7) can be offered simply in terms of the relative movement between neighbouring planes of atoms in the crystal structure caused by the deforming force. This point can be demonstrated very nicely if the student has access to a ball-raft model for the structure of crystals.† Elastic behaviour occurs where this relative movement is less than one atomic radius; clearly when a small deforming force is removed, the displaced raft will return to its original position. However, if a displacement in excess of one atomic radius occurs, the displaced raft will settle into a new position which corresponds to the permanent set for the sample strained just beyond the elastic limit. Plastic flow, on this model, is represented by the displaced raft slipping over a distance of several atomic radii. There is a limit to the amount of slip that can be produced in a given sample; this limit occurs because of the grain structure of polycrystalline materials and because of the various types of defect that can occur in the crystal structure.

Fig. 9:8 Typical load/extension graph for a material such as hard steel.

Plastic flow can be demonstrated by clamping a length of copper wire in a vice, gripping the other end tightly in a pair of pliers, and pulling it firmly; the yielding can be felt quite easily. However, the student is warned to brace him/herself suitably against the possibility of the wire breaking. Plastic flow is followed by work hardening. The material may become brittle before it actually fractures as in, say, the breaking of a piece of wire by repeatedly working it backwards and forwards until it finally snaps. The student might usefully consider why it is that "strong men" in circus acts bend metal bars, six-inch nails and the like from straight, rather than attempt to straighten them from the bent condition.

† A ball-raft may be constructed using table-tennis balls or commercially available polystyrene spheres. The spheres are stuck together in either a square or hexagonal close-packed arrangement in a plane. The crystal structure can be displayed by stacking two or three such rafts one on top of the other. Stable structures can be achieved when the convex surfaces of the spheres in an upper layer nestle comfortably into the gaps between spheres in a lower layer.

Just before the point of fracture occurs, surface cracks start to penetrate into the body of the sample, thus reducing the effective cross-section area and giving rise to the maximum stress. In some cases a "neck" forms at a point of weakness in the sample. Once these conditions have become established, fracture follows almost immediately.

9:6 Measurement of Young's Modulus for Steel

Young's modulus can be calculated directly from the slope of the linear portion, OP, of fig. 9:6. This involves the measurement of the unstrained length L, and the diameter d, of the wire. The length of the wire can be measured to an acceptable level of accuracy with a metre-rule; a micrometer gauge is used to measure the diameter.

The remaining observation that has to be made is the extension x; this must be made to the best attainable accuracy. Probably the simplest way to do this is as shown in fig. 9:9, where two similar wires are mounted side by side, one of which is the wire under test and the other is a reference wire. The reference wire carries a load which is just sufficient to keep it vertical and taut, while the test wire can be loaded up to about 6 kg in 0.5 kg steps. At its lower end, the reference wire carries a scale against which a vernier mounted on the test wire is able to move, thus enabling the extension to be measured to an accuracy of 0.01 mm. A slightly more elegant device for measuring the extension is Searle's extensometer, the essential features of which are shown in fig. 9:10.

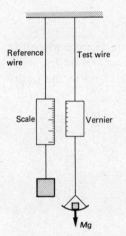

Fig. 9:9 Vernier and scale arrangement for measurement of the extension.

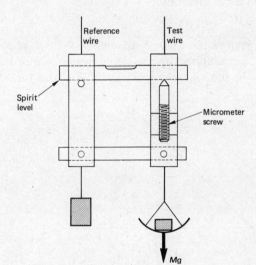

Fig. 9:10 Searle's extensometer. As load M increases, the spirit level is kept horizontal by adjusting the micrometer screw. The micrometer movement is equal to the extension of the wire.

It is important that measurements of the extension are made as the load is increasing and as it is decreasing. This provides a two-fold advantage; firstly it gives a repeated reading for the extension at each value of the load, and secondly it enables the observer to check whether the wire has been strained beyond the elastic limit. Good agreement between the two readings for each value of the extension indicates that there is no evidence of permanent set and thus it may be concluded that the elastic limit was not exceeded during the experiment. A load/extension graph of the type shown in fig. 9:6 should be plotted and the slope obtained in kg/m. Using the definitions for stress, strain and Young's modulus we obtain:

$$\text{Stress} = \frac{Mg}{\pi(\frac{1}{2}d)^2} = \frac{4Mg}{\pi d^2}\,\text{N/m}^2$$

$$\text{Strain} = \frac{x}{L}$$

$$E = \frac{\text{Stress}}{\text{Strain}} = \frac{4MgL}{\pi d^2 x}\,\text{N/m}^2$$

But the slope of the load/extension graph gives the average value for M/x. Thus

$$E = \frac{4gL}{\pi d^2} \times (\text{Slope})\,\text{N/m}^2 \qquad (9:5)$$

9:7 The Work Done in Stretching Wire

In the elastic deformation, the load increases from zero to its maximum value M^* and the extension from zero to x^* as shown in fig. 9:11. If we consider a small element of extension Δx when the load has a value M kg, then the work done by the force Mg newtons acting over the distance Δx is given by

$$\Delta W = Mg\,\Delta x$$

The area of the shaded strip in fig. 9:11 is $M\,\Delta x$ kg m. Hence, when this area is multiplied by the acceleration of free fall g, we obtain the work done in the elemental extension Δx. The total work done in the entire

Fig. 9:11 Work done in an elastic deformation.

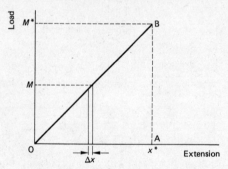

deformation will be given by the sum of all such strips, i.e. the total area OAB under the line *times* the acceleration of free fall. Thus

$$W = (\text{Area OAB}) \times g = \tfrac{1}{2}\text{OA} \times \text{AB} \times g = \tfrac{1}{2}x^* M^* g \qquad (9:6)$$

This expression for the total work done can be expressed in words as: one half of the product of deforming force and deformation. In physical terms, this result represents the work that has to be done against inter-atomic forces in the crystal structure of the wire during the deformation. It is sometimes referred to as the *energy stored in the strained wire*. This statement implies that this amount of work is recoverable when the wire returns to its unstrained condition. It is this stored elastic energy that is used to impart kinetic energy to the projectile fired from a catapult.

An interesting query arises when we consider that, when the load M^* is put on the wire and produces the extension x^*, it loses an amount of potential energy equal to M^*gx^*. This loss of potential energy supplies the work needed to deform the wire, which is only one-half of that which is available. The problem is how to account for the balance. A more advanced treatment than we can attempt here shows that, when the wire is stretched, its temperature drops slightly and consequently energy is required to bring the sample back to the temperature of the surroundings.

Example 9:2

A cylindrical metal bar of radius 1 cm and length 0.2 m is heated to a high temperature, firmly clamped and then allowed to cool. As a result the bar is under tension. If Young's modulus for the metal is $2 \times 10^{11}\,\text{N/m}^2$ and the strain is 3×10^{-3} determine the tension in the bar and calculate the energy stored in the bar.

Solution From equation (9:4) Young's modulus $= \dfrac{\text{stress}}{\text{strain}}$

so that $\quad 2 \times 10^{11} = \dfrac{\text{stress}}{3 \times 10^{-3}}$

$$\text{Stress} = 2 \times 10^{11} \times 3 \times 10^{-3} = 6 \times 10^{8}\,\text{N/m}^2$$

The area of cross-section of the bar is

$$\pi r^2 = \pi \times (10^{-2})^2$$

and so from equation (9:2) the tension is

$$6 \times 10^{8} \times \pi \times 10^{-4} = 1.88 \times 10^{5}\,\text{N}$$

To calculate the energy stored we need first to calculate the extension, which can be found from the strain. Substituting in equation (9:1) gives

$$3 \times 10^{-3} = \frac{x}{0.2}$$

$$x = 3 \times 10^{-3} \times 0.2 = 6 \times 10^{-4}\,\text{m}$$

From equation (9:6)

Energy stored $= \frac{1}{2} \times \text{force} \times \text{extension}$

Energy stored $= \frac{1}{2} \times 1.88 \times 10^{5} \times 6 \times 10^{-4} = 56.4\,\text{J}$

Exercises

Where necessary, take the acceleration of free fall $g = 9.8\,\text{m/s}^2$.

9.1 A steel wire of length 1.5 m and diameter 0.6 mm carries a load which is just sufficient to maintain it vertical and taut. Calculate the extension that is likely to be produced when a further load of 4.0 kg is added to the system. Young's modulus for steel $= 2.00 \times 10^{11}\,\text{N/m}^2$.

9.2 A steel wire of length 3.0 m hangs vertically from a rigid support and carries a load of 8.0 kg at its lower end. If the wire is 1 mm in diameter, and if the value of Young's modulus for steel is $200\,\text{GN/m}^2$, calculate *a*) the stress in the wire; *b*) the strain, assuming that the elastic limit has not been exceeded; *c*) the amount by which the wire extends as a result of the applied load; *d*) the energy stored in the stretched wire.

9.3 A uniform bar of length 0.4 m and of mass 10 kg is suspended from a rigid support by two vertical wires each of unstrained length 1.20 m and diameter 0.8 mm attached to its ends. Given that one wire is steel and the other is phosphor-bronze, calculate the angle of inclination of the bar to the horizontal when it hangs in equilibrium.

The values of Young's modulus for steel and phosphor-bronze are 200 GN/m^2 and 120 GN/m^2 respectively.

9.4 A steel tyre is to be fitted on to a wheel of radius 0.2 m. It is found that, when the tyre has been heated to 520 °C, it can just be slipped on to the wheel. Given that the cross-section area of the tyre is 5×10^{-4} m^2 and that Young's modulus for steel is 200 GN/m^2, estimate the tension in the tyre when it has cooled to 20°C, assuming that the elastic limit has not been exceeded. The value for the linear expansion coefficient of steel is 12×10^{-6} per degC.

9.5 In a certain building framework, a steel strut of length 2.0 m and cross-section area 10^{-4} m^2 is found to be experiencing a compression of 1000 N. Calculate the contraction it will show.

9.6 A catapult is constructed from a piece of rubber cord of total length 0.3 m and cross-section area 25×10^{-6} m^2. Given that the value of Young's modulus for the rubber used is 8×10^8 N/m^2, estimate a value for the velocity with which a projectile of mass 0.1 kg will leave the catapult if the rubber cord is extended by an amount equal to one-quarter of its natural length. What assumptions must be made in order to obtain this result?

9.7 In an experiment to determine Young's modulus for the material of a wire, a student obtained the following results:

Unstrained length of the sample = 1.66 m
Average value for the diameter of cross-section = 0.459 mm

| | Extension/mm | |
Load/kg	Load increasing	Load decreasing
0.00	0.00	0.00
0.50	0.33	0.31
1.00	0.68	0.64
1.50	0.97	0.93
2.00	1.26	1.24
2.50	1.58	1.56
3.00	1.90	1.88
3.50	2.17	2.18
4.00	2.49	2.49

Draw a suitable graph to display these results and obtain a value for Young's modulus of the material of the wire.

9.8 A long wire of length *L*, diameter 2*r*, of material of Young's modulus *E* hangs vertically in tension due to a load *M* and experiences an extension *x*. Assuming that the volume of the wire remains constant during the extension process, show that the radius of cross-section changes by an amount Δr, where

$$\Delta r = -\tfrac{1}{2}xr/L = -\tfrac{1}{2}Mg/\pi Er$$

Use the data given in 9.7 to estimate the change in the radius of the wire at maximum load.

10 Heat, Temperature and Internal Energy

10:1 Heat

In Chapter 4 energy was discussed with particular reference to its mechanical forms, and the important point was developed that, in certain cases, potential energy can be completely converted into kinetic energy. However, more generally, if we wish to speak of the conservation of energy it is necessary to consider various other forms in which energy can occur. Furthermore, it is often rather difficult to produce an accurate balance-sheet to account for these forms other than potential and kinetic energy. To illustrate this point, it may be helpful to look back to example 4:6 in which a high-velocity bullet becomes embedded in a block of wood. The calculated kinetic energies before and after the impact show that nearly 99% of the kinetic energy has been "lost" in the impact. This loss of kinetic energy can be explained if we take account of the heat generated by friction and the work done against intermolecular forces in the wood as the bullet penetrates the block. Neither of these quantities would be particularly easy to calculate, but it can safely be stated that the greater part of the lost energy is concerned with the conversion of kinetic energy into heat. In this chapter, the principal aim is to consider **heat as a form of energy**.

"Heat" is one of a number of words which, unfortunately, have become part of the tradition of physics but which lead to a certain amount of confusion and misunderstanding. In general, the use of the word heat as a noun adds nothing to our understanding of thermal physics; "heating" as a process may perhaps reasonably be taken to mean supplying energy to a sample. In order to clear up some of the confusion associated with the word heat it will be helpful to look briefly at the ideas of the kinetic theory of matter.

10:2 The Kinetic Theory of Matter

The simple **kinetic theory of matter** seeks to explain observed physical properties of solids, liquids and gases in terms of the motion of individual molecules. It must be stressed straightaway that the molecules of this model are highly idealised and should not be thought of as having any similarity to the molecules of chemical compound formation.

In **gases**, the molecules are considered to be in a state of random motion. This motion takes place throughout the whole volume within which the molecules are contained. At moderate pressures the molecules are sufficiently widely separated for them to have no appreciable effect on each other. Thus we may consider that the energy of the assembly of molecules is given by the sum of the individual kinetic energies. When energy is supplied to the sample, in the process called "heating", the kinetic energies of the individual molecules will increase. In order to introduce the idea of the *temperature* of the sample it is necessary to think of the *average kinetic energy* of all the molecules in the sample. On the kinetic model, the **absolute temperature** (see section 10:3) is defined as being proportional to the average kinetic energy of the molecules. Since the heating process leads to an increase in the kinetic energies of the individual molecules, it is clear that an increase in temperature will occur.

The total energy of the molecules in a sample is usually referred to as the **internal energy**. Thus in gases, the internal energy is wholly kinetic.

However, when we consider **solids** the situation is very different. Here the molecules are restricted to well-defined sites in the crystal lattice and the only form of motion that is available to them is *vibration*. In addition, the molecules are packed together in very close proximity to each other; this means that they are able to exert strong intermolecular forces on their near neighbours. Consequently, in a solid, the internal energy is made up of the vibrational kinetic energy plus a potential energy contribution from the action of the intermolecular forces. When a solid is heated, the energy supplied goes to increase the vibrational kinetic energy of the molecules; hence again an increase in temperature is to be expected. Eventually a point will be reached at which the vibrational energy becomes sufficiently great to start to break down the bonds between molecules; this means that the solid structure is lost and the solid melts.

Liquids are much more difficult to describe than either solids, where the structure is *highly ordered*, or gases, where the motion of the molecules is *completely random*. Indeed, the theory of liquids still continues to present a number of highly challenging problems to the researchers. When a solid melts, an increase in volume usually takes place; in this respect, water is a peculiar substance in so far as ice at the normal melting point has a lower density than liquid water. This change in volume on melting is only of the order of 10%; thus the separation between molecules in the liquid is not very different from that in the solid. This means that in liquids, the internal energy still contains a significant potential energy contribution, but the kinetic energy is now translational (i.e. can be represented by the familiar $\frac{1}{2}mv^2$) rather than vibrational. When a liquid is heated, we can think of the energy supplied as going largely to increase the kinetic energy of the molecules, as the temperature rises to the boiling point.

We shall make considerable use of this simple kinetic model of matter in the next two chapters.

10:3 Scales of Temperature

In the kinetic model, we have used the idea of temperature to gauge the internal energy of the sample; this is sometimes referred to as the *hotness* of the body. In order to make comparisons between different samples it is necessary to use a **temperature scale**. Probably the very first temperature scale consisted of just three points: "too hot", "too cold" and "just right". This may be appropriate for judging the temperature of porridge or the baby's bath water but it is hardly a scientific scale. In order to make a temperature scale quantitative, various *fixed points* are needed; these should be easily achievable and accurately reproducible. It is convenient to make the **lower fixed point** that at which melting ice and pure water are in equilibrium, and the **upper fixed point** that at which pure water boils. However, the boiling point and the freezing point of water both depend on the pressure in the system, so it is also necessary to specify that both of these points are fixed at *standard atmospheric pressure* (1.013×10^5 Pa). **Celsius** temperatures of 0°C and 100°C are associated with the lower and upper fixed points respectively, which are usually known as the *ice point* and the *steam point*.

For practical temperature measurement it is necessary to make use of some physical property of matter whose magnitude varies with temperature. Examples of various physical properties used in this way are:

a) the length of a liquid column;

b) the pressure of a gas at constant volume;

c) the resistance of an electrical conductor;

d) the thermo-electric (Seebeck) effect;

e) the colour of an incandescent object;

f) the saturated vapour pressure.

This is by no means a complete list and various rather "exotic" techniques may be satisfactorily employed especially in the very low and very high temperature regions. These are essentially specialist methods and will not be dealt with here.

Let X_t denote the magnitude of some physical property (e.g. the length of a mercury thread in a capillary tube, or the resistance of a length of platinum wire) at some temperature t°C. Then X_0 will denote the magnitude of the property at the ice point and X_{100} that at the steam point. The difference $X_{100} - X_0$ is usually referred to as the *fundamental interval*. Celsius temperatures are specified in terms of **degrees** which are one-hundredth of the fundamental interval. Thus the temperature t°C can be defined by the equation:

$$\frac{t}{100} = \frac{X_t - X_0}{X_{100} - X_0} \tag{10:1}$$

It is necessary to mention that the celsius scale is not the only one which is used. In the previous section, where the kinetic model of gases was mentioned, it was stated that the average kinetic energy of the gas molecules comprising the sample was proportional to the *absolute temperature*. This scale of temperature is sometimes also known as the **kelvin scale**, or the *absolute thermodynamic scale*. A detailed discussion of this scale is beyond the scope of this book, but it is sufficient to state that, while the kelvin scale uses different fixed points from the celsius scale, the kelvin values are allocated in such a way that the fundamental interval between the ice and steam points remains 100 degrees.

The working rule is to add 273 (strictly, 273.15) to the celsius temperature in order to obtain the absolute temperature in kelvin. Thus the kelvin temperature for the melting point of pure ice is 273.15 K and for the boiling point of pure water is 373.15 K. In all but the most precise work the small decimal fraction (0.15) can be ignored.

Example 10:1

Given that the resistance of a coil of platinum wire is found to be 5.234 Ω when it is immersed in a beaker of melting ice, and 6.849 Ω when it is immersed in the steam just above the surface of boiling water, calculate a value for the temperature of the air in the laboratory if the resistance of the wire measured in air is found to be 5.557 Ω.

Solution $X_t - X_0 = 5.557 - 5.234 = 0.323 \ \Omega$
and $X_{100} - X_0 = 6.849 - 5.234 = 1.615 \ \Omega$

Hence in equation (10:1) $\dfrac{t}{100} = \dfrac{X_t - X_0}{X_{100} - X_0} = \dfrac{0.323}{1.615}$

$$t = \frac{0.323 \times 100}{1.615} = 20°C$$

10:4 Practical Temperature Measurement— Mercury Thermometer

For much temperature measurement, particularly in teaching laboratories, temperatures can conveniently be measured by use of the well-known **mercury-in-glass thermometer** calibrated in celsius degrees. The advantages of this instrument are that it is *a*) simple and inexpensive, *b*) convenient to use, *c*) reliable, and *d*) applicable over a fairly wide range of temperatures: −39°C to 360°C.

Mercury is chosen as the thermometric substance here because, being a liquid metal, it is a good thermal conductor and has a large expansion coefficient which is uniform over the temperature range for which it is used. By containing the mercury sample in a thin-walled bulb connected to a fine-bore capillary tube, it is possible to achieve quite large changes in the length of the mercury thread in the capillary for relatively modest temperature changes in the system with which the bulb is in contact. As is the case with all thermometers, it is important that good thermal contact should be achieved between the thermometer bulb and the object whose temperature is to be measured. This ensures that the thermometer and the object come to *thermal equilibrium* (i.e. they come to the same temperature) as quickly as possible. This process of achieving thermal equilibrium requires that there shall be some exchange of energy between the thermometer and the object; in general this exchange is small and can be neglected.

10:5 The Platinum Resistance Thermometer

The thermometer "bulb", in a **platinum resistance thermometer**, consists of a coil of pure platinum wire which is wound (non-inductively) on an insulating former. The coil should show an increase in resistance of about 40% between the ice and steam points. The free ends of the platinum coil are welded to base metal leads at the points A and B, as shown in fig. 10.1(a). A pair of compensating (or "dummy") leads is also included; these are identical to the thermometer leads and, as is clear from fig.

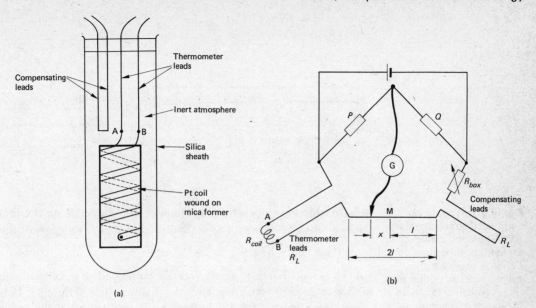

(a)

(b)

Fig. 10:1 (a) Schematic representation of the "bulb" of a platinum resistance thermometer (b) Callendar-Griffiths bridge

10:1(b), enable temperature-dependent effects in the leads to be compensated. The whole arrangement of coil, former and leads is enclosed in an inert atmosphere inside a silica sheath. The thermometer is used in a modified form of the Wheatstone bridge network and is shown schematically in fig. 10.1(b). The decade box provides "coarse" adjustment and the slide wire "fine" adjustment in the balance.

Because of the nature of its construction and the physical size of the bulb, the platinum resistance thermometer extracts a significant amount of energy from the source, and also exhibits a time-lag in coming to thermal equilibrium with the object. It is these considerations that make this device unsuitable for use with a) small samples and b) conditions where the temperature is changing. The circuit permits the measurement of resistance and hence steady temperatures to a high degree of accuracy.

10:6 The Thermocouple

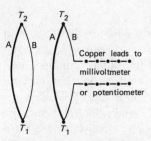

Fig. 10:2 Thermoelectric circuit to show Seebeck effect.

If wires of two dissimilar metals are joined together to form a closed loop, and the junctions are maintained at different temperatures, it is observed that an e.m.f. is set up in the circuit (see fig. 10:2). This phenomenon is known as the *Seebeck effect*; its explanation is an interesting topic in Solid State Physics but one on which we shall not embark here.

Since the e.m.f. developed depends on the temperature difference between the junctions, we have the basis of a very convenient means of temperature measurement since potentiometric methods are readily available for the measurement of even very small e.m.f.s. It is necessary to ensure that one junction is maintained at some suitable reference temperature (e.g. 0°C) and that the small e.m.f. is measured as accurately as possible. In general, the e.m.f. generated is only of the order of millivolts when the temperature difference between the two junctions is 100°C. The

101

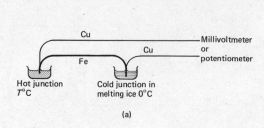

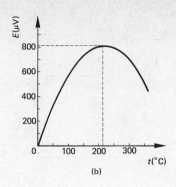

(a)

(b)

Fig. 10:3 (a) Typical experimental arrangement for a Cu/Fe thermocouple (b) Emf/temperature curve for Cu/Fe. (The results presented here were obtained on an experimental set-up in the authors' laboratories.)

relationship between e.m.f. and temperature depends on the metals used, and can usually be adequately described by a parabolic law of the form:

$$E_{AB} = aT + bT^2$$

where A and B relate to the metals employed (e.g. copper and iron), a and b are constants, and T is the temperature difference between the junctions. Clearly, if the *cold junction* is maintained at the ice point then T gives the celsius temperature of the *hot junction*. Fig. 10:3(a) shows a convenient arrangement for such a system, and fig. 10:3(b) depicts the e.m.f./temperature response for a Cu/Fe **thermocouple**. From fig. 10:3(b) it can be seen that Cu/Fe does not provide a very reliable thermometer for temperatures above about 180°C, as the e.m.f. approaches its maximum value and only changes slowly with temperature in this region.

One common example of metals used in thermocouples is chromel/alumel which shows a very good approximation to a linear temperature/e.m.f. response over a wide temperature range (−196°C to 1000°C).

The thermocouple is particularly suited to the measurement of changing temperatures or to the measurement of the temperature at a localised point in a sample. These uses come about because of the small physical size of the hot junction and the fact that the junction absorbs only a very small amount of energy from the object in order to come to thermal equilibrium. The nature of this instrument also makes it very suitable for use in an evacuated chamber where electrical leads can be taken easily through the walls of the chamber by specially insulated conductors.

10:7 Heat Capacity

It is a common fact of experience that when energy is supplied to a solid or liquid sample the temperature of the sample rises, provided that the solid remains solid (i.e. it does not melt) or that the liquid remains liquid (i.e. it does not evaporate or boil). This is a situation that can be investigated quite simply under laboratory conditions. Fig. 10.4 shows a Dewar flask containing a quantity of water; a small electric heater of known power rating can be used to supply energy at a constant rate for a known time and the temperature rise is then measured. Suppose that the heater is rated at 25 W and that the flask initially contained 0.2 kg of

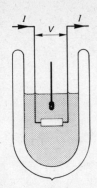

Fig. 10:4 A simple laboratory set-up to demonstrate heat capacity measurements.

Fig. 10:5 Temperature variation of the specific heat capacity of water. The results shown here are based on those obtained by Osborne, Stimsom and Ginnings in a classic experiment.

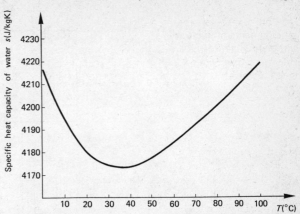

water at room temperature (20°C approximately). If the heater were run for 5 minutes (300 s) the energy supplied would be 7500 J and a temperature rise of about 9°C would be observed. If the experiment were repeated with say 0.5 kg of water, a temperature rise of around 3.6°C would be achieved in a 5-minute period. This would suggest that, for a given input of energy, the temperature rise achieved in the water was inversely proportional to the mass of water present. Alternatively, the experiment could be performed in such a way as to relate the observed rise in temperature of a given mass of water to the energy input. No matter how the experiment were arranged, the results would be consistent with the relationship:

$$\frac{Q}{m(T - T_0)} = s = \text{constant} \tag{10:2}$$

where Q denotes the energy input (i.e. the increase in internal energy) to a sample of mass m, producing a temperature change from an initial value T_0 to a final value T. The constant s is known as the **specific heat capacity** of the material of the sample; it will have units J/kg K for Q in joules, m in kilograms and the temperatures measured in either kelvins or celsius degrees. The product of the mass and the specific heat capacity, ms, is usually known as the **thermal capacity** or **heat capacity** of the sample.

A more precise analysis of this experiment would require us to take account of the energy taken up by the Dewar flask itself; this is only a very small quantity and would not affect the observations significantly under the conditions of a teaching-laboratory demonstration experiment. Again under very precise conditions, it could be shown that the observed value for the specific heat capacity of a substance depends on the temperature. Indeed, water is a substance which shows quite a marked variation of specific heat capacity with temperature; at 0°C the value of the specific heat capacity is about 4220 J/kg K but at room temperature (20°C) the value is only 4180 J/kg K. As the temperature is further increased, a minimum value of about 4175 J/kg K is observed in the region of 30°C and at 100°C the value has returned approximately to 4220 J/kg K (see fig. 10:5).

103

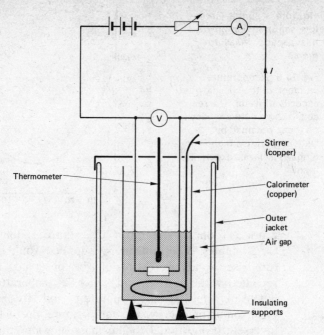

Specific heat capacities are usually **measured electrically**, but sometimes an archaic method known as *method of mixtures calorimetry* is employed. We shall describe briefly the electrical method. Fig. 10.6 shows a copper calorimeter inside an insulating jacket which prevents energy exchange with the surroundings; insulation is provided either by a layer of felt lagging or simply by an air gap. The calorimeter is a copper can, usually of mass in the range 0.1 to 0.2 kg, and the specific heat capacity of copper is assumed to be known—either from a previous experiment or by reference to tables. The calorimeter contains a measured mass of a liquid, whose specific heat capacity it is desired to determine. The electrical circuit supplies energy at a measured rate VI watts and the temperature rise obtained over a time interval t seconds is observed. In order to minimise heat losses it is usual to pre-cool the liquid to say 5°C below room temperature and to continue the heating process until the temperature has risen to 5°C above room temperature. A small hand stirrer is employed to ensure a uniform temperature distribution throughout the whole liquid sample.

The energy provided by the heater is VIt joules and, assuming that none of this energy is lost to the surroundings, we may equate this to the energy taken in by the calorimeter and its contents and hence calculate the specific heat capacity of the liquid. This calculation is illustrated in example 10:2.

A rather more sophisticated variant of this experiment is provided by *Callendar & Barnes' method of continuous flow*. Here, it is necessary that the liquid should be available in quantity, hence water is often used. This flows at a steady rate over an electrical heater; observations of temperatures and the flow rate are only made when steady temperatures

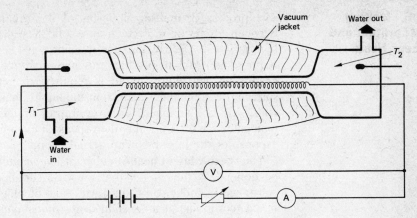

Fig. 10:7 Schematic arrangement of the apparatus for Callendar and Barnes continuous flow method for the specific heat capacity of water.

at the input and output points are recorded. In order to minimise the effects of energy losses from the system, the parameters are changed, and a second run is performed with different values for the power supply and the flow rate which will lead to the same values of the input and output temperatures. In practice, this is a fairly difficult condition to satisfy in a student's experiment where perhaps no more than $1\frac{1}{2}$ hours is available. Fig. 10:7 shows the experimental layout.

Example 10:2

A copper calorimeter of mass 50 g and specific heat capacity 390 J/kg K contains 35 g of alcohol which is initially at 15°C. The liquid is heated by an electrical heater with a potential difference of 3 V across it and a current of 0.5 A passes for 10 minutes (600 sec). The final temperature of the alcohol is 24°C. Assuming no energy is lost to the surroundings calculate the specific heat capacity of the alcohol.

Solution   Total energy supplied by the heater is

$$VIt = 3 \times 0.5 \times 600 \text{ J} = 900 \text{ J}$$

Energy taken in by calorimeter can be calculated from equation (10:2):

$$Q = ms(T - T_0) = 50 \times 10^{-3} \times 390(24 - 15) = 175.5 \text{ J}$$

Energy taken in by the liquid, using the same equation, is

$$35 \times 10^{-3} \times s \times (24 - 15) = 0.315s \text{ J}$$

If no energy is lost from the system we can apply the conservation of energy and equate the energy supplied to the energy taken in by the calorimeter and contents, i.e.

$$900 = 175.5 + 0.315s$$

$$0.315s = 900 - 175.5 = 724.5$$

$$s = \frac{724.5}{0.315} = 2300 \text{ J/kg K}$$

105

10:8 Melting, Evaporation and Latent Heats

The process of **melting** in terms of the breaking down of the bonds between atoms or molecules in a solid has already been mentioned in section 10.2. Clearly, an energy input at the melting temperature is required in order to produce this *phase change*. Similarly when a liquid **evaporates**, an energy input is required to get molecules through the surface, i.e. to change them from the liquid phase to the vapour phase. If energy is not supplied, an evaporating liquid is observed to cool as molecules are lost; we may then speak of evaporation taking place at the expense of the liquid's own internal energy.

The **specific latent heat** of either *fusion* (*melting*) or of *evaporation* may be defined as the energy that must be supplied to unit mass (1 kg) of substance in order to produce a change of phase (i.e. solid to liquid, or liquid to vapour) at a *constant temperature*. We shall illustrate this point by the consideration of a worked example.

Example 10:3

Water in an unlagged boiler is heated electrically. When the water is boiling steadily, the steam evolved is passed through a condenser and is collected in a pre weighed container. In two successive runs the following observations of the potential difference across, and current in, the heating element were recorded:

a) 100 volts, 6.60 amps, 27.8×10^{-3} kg collected in 2 mins

b) 50 volts, 4.10 amps, 3.48×10^{-3} kg collected in 2 mins.

Use these data to determine a value for the specific latent heat of vaporisation of water at the normal boiling point, and the power loss from the boiler at this temperature.

Solution When the water is boiling steadily some of the power supplied goes to converting water from liquid to vapour but the rest is effectively "lost" from the apparatus. We may think of this lost power as maintaining the apparatus at the steady boiling temperature. Suppose that this is H J/s.

Hence, in general, when the power supply is VI watts and m kg/s of water are being vaporised we may write:

$$VI = mL + H \tag{10:3}$$

Thus for run a) $100 \times 6.60 = \left(\dfrac{27.80 \times 10^{-3}}{120} L \right) + H$

and for run b) $50 \times 4.10 = \left(\dfrac{3.48 \times 10^{-3}}{120} L \right) + H$

$\therefore \quad 660 = (2.317 \times 10^{-4} L) + H$

and $205 = (0.290 \times 10^{-4} L) + H$

Hence, subtracting to eliminate H $455 = 2.027 \times 10^{-4} L$

$$L = \frac{455}{2.027} \times 10^4 = 2245 \times 10^3 \text{ J/kg}$$

In order to obtain a value for H, it is necessary to substitute this value for L into one of the above equations:

$660 = (2.317 \times 10^{-4} \times 2245 \times 10^3) + H$

$660 = 520.2 + H$ i.e. $H = 139.8$ W

This means that if the power supply were to be reduced to 139.8 W it would be just sufficient to maintain the system at the normal boiling point.

All the power supplied goes to making good the energy loss to the surroundings and there would be no energy available to produce vaporisation. The system would be just maintaining water as liquid at its normal boiling temperature.

Fig. 10:8 Self-jacketing vaporiser to determine the specific latent heat of vaporisation

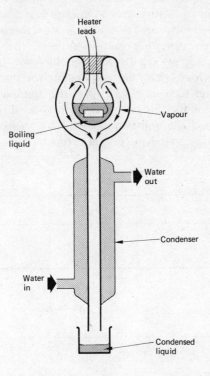

Fig. 10:8 shows an experimental arrangement used for measuring the latent heat of vaporisation which is designed to reduce the energy loss term H in equation (10:3). The vapour boiled off surrounds the boiling liquid and so insulates it from the surroundings. Measurements are made once steady state conditions have been achieved and the calculation proceeds as in example 10:3.

10.9 Heat Transfer

The three principal methods of heat transfer are *conduction*, *convection* and *radiation*; we shall deal briefly with these. In the previous chapter, some ideas of the structure and crystalline nature of solids were introduced. Since the atoms (ions or molecules) are bound to lattice sites, the only form of motion that is permitted to these atoms is vibration about the lattice point. We could, therefore, say that the internal energy of a sample of a solid material was given by the sum of the potential energies, due to the electrostatic nature of the bonding within the lattice, and the vibrational kinetic energies. Consider now a long bar of a crystalline solid material with one end heated, either electrically or by a steam chest, and the other end cooled in some way. At the hot end, the energy supplied by the heater goes to increasing the kinetic energy of the atoms; this shows up as an increase in the amplitude of the vibrations. Hence the internal energy and, therefore, the temperature are raised locally in the bar. However, since the atoms in the lattice are coupled by the forces giving rise to the bonding, this increase in amplitude is handed on to neighbouring atoms in the lattice and so energy is gradually transmitted along the bar. This is the process that we know as **thermal conduction**.

The heat transfer mechanism through the lattice that has been described applies principally to *insulators* and non-metals. In metals the "free" electrons which are responsible for the process of electrical conduction (refer to section 13.2) play the dominant role in thermal conduction also. It is for this reason that, in general, good electrical conductors are also good thermal conductors. This process we have described for solids can be applied in general terms to liquids and gases also.

Fig. 10:9 Schematic representation of situation for the definition of thermal conductivity, considering heat flow through a section of a rectangular bar.

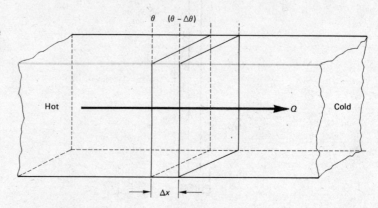

The **thermal conductivity** for a slab of material of thickness Δx and with faces of area A is defined in terms of the rate of flow of energy between the faces and the temperature difference $\Delta\theta$ across the slab (see fig. 10:9). The defining equation for the thermal conductivity is

$$Q = -kA\frac{\Delta\theta}{\Delta x} \tag{10.4}$$

where Q denotes the energy flowing per second through the slab. The units for thermal conductivity are W/m K. Methods for the measurement of thermal conductivity depend upon whether the sample may be classed as a "good" or a "poor" conductor. A standard method for good conductors is known as *Searle's bar*; we shall consider this by means of a worked example.

Example 10:4

A copper bar of diameter 25 mm has one end brazed into a steam chest and the other end has two or three turns of narrow-bore copper tubing brazed on to it. The whole arrangement is enclosed in thick felt lagging. Two thermometers are used to measure temperatures at points 100 mm apart in the central region of the

Fig. 10:10 Schematic arrangement of Searle's bar experiment for measurement of the thermal conductivity of a good conductor.

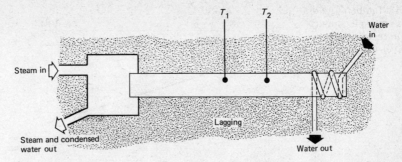

bar. Cold water flows at a constant rate through the copper tubing and the in-flow and out-flow temperatures are measured (see fig 10:10). Using this arrangement, temperatures of 65.2°C and 48.7°C were recorded by the thermometers T_1 and T_2 when steady conditions were established; when 0.1 kg of water was observed to flow through the copper tube in 51.0 sec, in-flow and out-flow temperatures of 14.7°C and 18.5°C respectively were recorded. Given that the specific heat capacity of water is 4180 J/kg K obtain a value for the thermal conductivity of copper.

Solution Since the bar is well-lagged and copper is a good conductor, it is reasonable to assume that all the energy which enters the bar at the hot end flows uniformly along the bar and is taken up by the cooling water at the cold end. (To use an electrical analogy, we are saying that the "heat current" along the bar is constant.)

We may write for the temperature gradient in the centre of the bar:

$$\Delta\theta/\Delta x = (48\cdot7 - 65\cdot2)/0.1 = -165 \text{ K/m}$$

We may calculate the heat current from the temperature rise in the cooling water. The flow rate for the water is 0.1/51 kg/s; hence the rate of take-up of energy by the water is

$$(0.1/51)4180(18.5 - 14.7) = 31.1 \text{ J/s}$$

We may now apply this heat current to the readings obtained in the central portion of the bar, using eqn. (10.4):

$$31.1 = -k\pi(12.5 \times 10^{-3})^2 \times (-165)$$

giving a value for $k = 384$ W/m K.

If the bar in fig. 10:10 is not lagged, heat will be lost from the sides of the bar and so the heat current decreases along the bar. This results in a change in the temperature gradient along the bar as illustrated in fig. 10:11.

For composite bars the same general principle can be applied. If a copper bar is joined end-on to an iron bar we can consider the same heat current to flow in each part of the system, provided that it is well-lagged. Then the problem can be regarded as analogous to that of the flow of electric current through two resistors connected in series. These ideas can be applied to practical cases such as single and doubled glazed windows, and walls which may or may not be plastered and which may have either an air cavity or foam-filled cavity between them. In practice, construction engineers tend to use the idea of *thermal transmittance*, or as it is more

109

Fig. 10:11 Temperature distributions along a conducting bar in cases where the bar is (a) well-lagged and (b) unlagged.

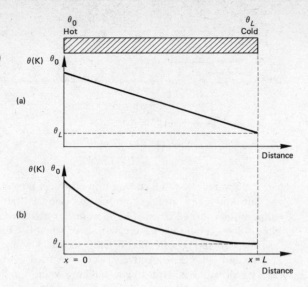

widely known the *U-value*, when considering the flow of heat through walls and roofs. This approach enables account to be taken of a variety of factors such as internal and external conditions appropriate to the building: moisture, exposure to rain, wind speed, etc.

In conduction we have described the process of heat transfer in terms of the handing on of energy from one atom or molecule to its neighbours through an increase in the vibrational energy; there is no bulk movement of the atoms in the sample. The process of **convection** does, however, allow for bulk movement. If a beaker of water is heated very gently, by allowing a small gas flame to play on a small area of the base for a short period of time, then the water is only heated very locally. Since water is a relatively poor conductor of heat, the temperature distribution in the bulk of the water is not significantly changed. The locally warmer water, however, expands and consequently experiences a local decrease in density. This warmer less-dense water rises and its place is taken by colder, denser water sinking down from the top of the beaker. Hence, a circulation within the bulk of the sample is set up; this motion of the water is known as a *convection current*. Convection currents can easily be demonstrated by introducing a small crystal of a marker dye (e.g. potassium permanganate) into the water (see fig 10:12).

Convection currents in gases, particularly air, are also of interest and importance. Space heating and ventilation in buildings depend upon the process of convection. In many laboratory experiments where the rate of loss of heat has to be taken into consideration, convection loss is the most significant feature when the temperature excess over the surroundings is not too great. It is sufficient for our purposes to state that for modest temperature differences between the sample and its surroundings, the rate of convective heat loss can be expressed as:

$$\text{Rate of loss of heat} \propto (\text{Excess temperature})^n$$

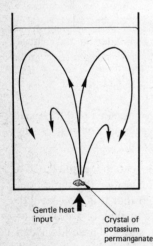

Gentle heat input

Crystal of potassium permanganate

Fig. 10:12 Convection currents (idealised) in a beaker of water subject to local heating.

The index n has the value 1 for *forced convection*, i.e. when the sample is suspended in a draught of air, and the value 1.25 for *free convection*, i.e. when careful precautions are taken to screen the sample from draughts.

The process of **radiation** has already been mentioned in an earlier chapter (see section 8:6 and fig. 8:8). The radiation process can be thought of as a means of heat transfer between a source and a detector without any appreciable heating of the space between them. Here, we are primarily concerned with the infra-red region of the electromagnetic spectrum. Heat radiation obeys all the laws which govern electromagnetic radiation.

Heat loss by radiation is most significant where there is a large temperature difference between the sample and the surroundings. Under these conditions the rate of radiative heat loss is proportional to the fourth power of the absolute temperature.

Exercises

10.1 Suggest an appropriate thermometer for use in each of the following temperature measurement situations; give reasons for your choice of thermometer in each case:

a) A thermal conductivity experiment involving a bar-shaped sample of copper, in which it is desired to measure two steady temperatures of about 70°C and 50°C to an accuracy of ±0.5°C.

b) An experiment to determine the freezing point of zinc (about 420°C) by plotting a cooling curve. The metal is available in quantity and the best attainable accuracy is required.

c) A heat capacity experiment involving a small sample of indium maintained in an evacuated enclosure at liquid nitrogen temperature (77 K). A temperature rise of about 5°C is to be produced in the sample by means of energy supplied from a small resistance heater wound directly on, but electrically insulated from, the sample.

10.2 A certain thermocouple in which one junction is maintained in melting ice is found to develop e.m.f.s of 4.09 mV and 13.36 mV when the hot junction is maintained at the temperatures of water boiling at normal atmospheric pressure and of the freezing point of lead (327°C) respectively. Use these data to estimate the freezing point of tin, given that the corresponding e.m.f. developed is 9.44 mV.

10.3 A platinum resistance thermometer is calibrated by immersing the bulb first in melting ice and then in the steam from boiling water; the values recorded for the resistance are 2.619 Ω and 3.618 Ω respectively. When the bulb is put in good thermal contact with a sample of solid carbon dioxide, the resistance is found to be 1.872 Ω. Use these data to estimate the equilibrium temperature of solid CO_2 with its vapour at normal atmospheric pressure.

10.4 It has been shown that for pure strain-free platinum the resistance varies with temperature according to the equation:

$$R = R_0(1 + \alpha\theta + \beta\theta^2)$$

where $\alpha = 4 \times 10^{-3}$ deg^{-1}, $\beta = -6 \times 10^{-7}$ deg^{-2}, R_0 is the resistance at 0°C and θ is the Celsius temperature measured with a gas thermometer. Obtain values for the

111

resistance of the thermometer element at gas thermometer temperatures of 50°C and 100°C. Determine what temperature the platinum resistance thermometer indicated when the gas thermometer temperature was 50°C.

10.5 In a certain experiment it was found that when a copper sample of mass 0.1 kg absorbed energy amounting to 195 J its temperature rose from 15°C to 20°C. Estimate the specific heat capacity of copper.

10.6 An empty copper calorimeter has a mass of 0.135 kg. When it was about half full of water, the total mass was found to be 0.404 kg. A separate piece of copper, of mass 0.112 kg, was heated to a temperature of 100°C and rapidly transferred to the water in the calorimeter. The mixture was carefully stirred and it was observed that the temperature of the water rose from 22.8°C to 25.7°C. Assuming that the system was well insulated to prevent energy exchange with the surroundings, and given that the specific heat capacity of water is 4180 J/kg K, obtain a value for the specific heat capacity of copper.

10.7 A metal block, of mass 0.444 kg, has a resistive heating element wound on it, but electrically insulated from it. In an experiment a potential difference of 11.6 V was maintained across the ends of the heater for a period of 10 minutes giving rise to a current of 0.6 A in the element. After applying an appropriate correction for energy losses, a temperature rise of 8.4°C was recorded. Obtain a value for the specific heat capacity of the metal.

10.8 In a continuous flow experiment to determine the specific heat capacity of a liquid (see fig. 10.7) constant in-flow and out-flow temperatures of 15.3°C and 20.7°C were maintained in two runs. In the first run the heater was operated at 12.0 V and 0.8 A and it was found that 49.4 g of liquid were collected in a period of 2 minutes. In the second run, the corresponding values were 7.5 V, 0.5 A and 43.2 g of liquid were collected in 5 minutes. Calculate the specific heat capacity of the liquid and the rate of energy loss from the device.

10.9 In an experiment to determine the specific latent heat of fusion of ice, in which small pieces of dried crushed ice were added to water in a well-lagged calorimeter, a student recorded the following results:

mass of empty calorimeter and stirrer	132.3 g
mass of calorimeter, stirrer and water	341.7 g
total mass after adding ice	364.1 g
initial temperature	20.0°C
final temperature	11.3°C

Given that the specific heat capacity of the material of the calorimeter was 400 J/kg K, and that the specific heat capacity of water is 4180 J/kg K, obtain a value for the specific latent heat of fusion of ice.

10.10 Using apparatus of the type illustrated in fig. 10.8, alcohol was boiled at normal pressure and the condensate collected in a pre-weighed vessel. In one run, when the potential difference across the heater was 20 V and the current 2.0 A, 10.1 g of condensate were collected in a period of 5 minutes; in a second run at 25 V and 2.5 A, 16.2 g of condensate were collected in a similar period. Calculate a value for the specific latent heat of vaporization of alcohol. Calculate also the rate of loss of energy from the system at this temperature.

10.11 A suitably insulated calorimeter of thermal capacity 126 J/K contains 0.2 kg of water at 20°C. 0.05 kg of clean dry ice at 0°C are added to the calorimeter and the mixture is carefully stirred. It was observed that the temperature of the calorimeter and its contents dropped to 2.4°C. Given that the specific heat capacity of water in this region may be taken as 4200 J/kg K, use these data to estimate a value for the specific latent heat of fusion of ice at 0°C.

10.12   Compare the rates of heat flow along a copper bar and an iron bar of the same dimensions when one end of each bar is maintained at the steam temperature and the other end at the ice temperature. Assume that in each case the bar is well lagged. The thermal conductivity of copper is 385 W/m K, and for iron it is 48.5 W/m K.

10.13   A copper bar of length 150 mm and diameter 10 mm is joined end-on to an iron bar of the same diameter but of length 100 mm. The system is in a well-lagged condition with the end of the copper bar maintained at 100°C and the end of the iron bar at 0°C. Using the values for thermal conductivities given in 10.12 estimate the rate of flow of heat in the composite bar and the temperature of the interface between the copper and iron.

10.14   A glass window pane of area 1 m^2 is made of glass 4 mm thick and of thermal conductivity 1 W/m K. Estimate the rate of heat flow through the pane if the inside air temperature is 20°C and the outside temperature is 5°C. If the window is double-glazed by mounting a similar sheet of glass in front of the first one to enclose an air gap of thickness 5 mm, obtain a value for the rate of heat flow in these circumstances. Assume the thermal conductivity of air is 25×10^{-3} W/m K, and that the inside and outside temperatures remain unchanged.

10.15   In an experiment to determine the energy dissipation in a candle flame, samples of water each of mass 0.200 kg and initially at 25°C were heated by the candle in a glass beaker of thermal capacity 100 J/K. A total of five runs was carried out, each of duration 5 minutes, and the following values were recorded for the final temperatures: 39.0°C, 40.0°C, 42.0°C, 42.0°C and 39.5°C. The initial mass of the candle was 35.1 g and after burning continuously for 36 minutes its mass was 31.3 g. Use this data to estimate the mean power supply provided by the candle, the time needed to completely burn the candle, and the total amount of energy it is capable of supplying. (Specific heat capacity of water = 4180 J/kg K.)

11 The Gaseous State

11:1 The Gas Laws

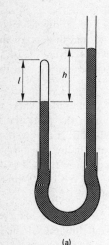

(a)

h(mmHg)

$\frac{1}{l}$(m⁻¹)

P

(b)

Fig. 11:1 Investigation of Boyle's law.
(a) shows a fixed mass of gas confined to length *l* of uniform cross-section. If the atmospheric pressure is *P* the total pressure is *P* + *h*.
(b) Plot of *h* (measuring pressure) against 1/*l* (which measures 1/*V*). Graph will be linear if Boyle's law holds. Extrapolation onto *h* axis gives atmospheric pressure *P*.

The study of the properties of gases, particularly air, is one which has been of interest from the beginnings of the history of science and technology. Much, if not all, of the early work contained little that we would recognise as "science" today but it did show great technological skill. By the second half of the 17th century the development of the mercury barometer by Torricelli and of the vacuum pump by van Guericke stimulated the scientific study of gases.

Around 1660, Robert Boyle proposed what we now recognise as probably the first genuine law of physics:

Boyle's Law
For a fixed mass of gas, the volume varies inversely with the pressure provided that the temperature remains constant.

i.e. $p \propto \dfrac{1}{V}$ or $p = \dfrac{k}{V}$ $\qquad$ (11:1)

where k is a constant. Figs. 11:1(a) and 11:1(b) show a simple laboratory set-up whereby this law may be demonstrated. Gas changes which take place at constant temperature are called **isothermal changes**.

Subsequent investigations led to what we now know as Charles' Law and the pressure law. These may be stated simply as follows:

Charles' Law For a fixed mass of gas maintained at constant pressure, the volume increases by a constant fraction of its volume at 0°C for every celsius degree rise in temperature:

$$V = V_0(1 + \alpha t) \qquad (11:2)$$

where α is known as the *volume coefficient of expansion*.

Pressure law For a fixed mass of gas maintained at constant volume, the pressure increases by a constant fraction of its pressure at 0°C for every celsius degree rise in temperature:

$$p = p_0(1 + \beta t) \qquad (11:3)$$

where β is known as the *pressure coefficient*.

It has been shown that within the limits of experimental accuracy:

$$\alpha = \beta = \frac{1}{273} \text{ per degree}$$

Fairly simple laboratory demonstrations may be performed to illustrate these two laws.

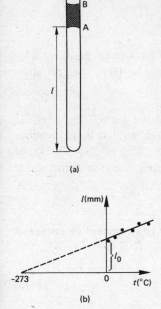

(a)

(b)

Fig. 11:2 Variation of gas volume with temperature. (a) shows length *l* of gas confined by plug AB. *l* is proportional to the gas volume. (b) Plot of variation of length *l* with temperature *t*. l_0 is the length at 0°C corresponding to volume V_0.

1 Fig. 11:2(a) shows how the volume variation may be investigated. A sample of dry air is contained in a length of uniform narrow-bore glass tubing; the air column is enclosed at the top by a short plug AB of concentrated sulphuric acid, which serves as a piston and also maintains the air dry. The plug is most conveniently introduced into the tube with a small glass pipette drawn out into a long fine nose; all the usual precautions for the handling of conc. H_2SO_4 should be observed. The tube is mounted against a millimeter scale with a thermometer so that it can be gently heated in a water bath over a range from about 4°C to 80°C. Since the bore of the tube is uniform, the length of the air column is directly proportional to the volume of the fixed mass of dry air trapped beneath the plug. A typical set of results for such an experiment is shown in fig. 11:2(b). If the graph is extrapolated back (i.e. produced backwards beyond the region where the experimental points were obtained), it can be shown that it cuts the temperature axis in the region of −273°C. Clearly the slope of the line is 1/273 per degree.

2 A simple form of laboratory constant-volume gas thermometer can be used to show the pressure variation with temperature as shown in fig. 11:3(a). The fixed mass of gas under consideration is contained in the bulb A and that part of the neck of the bulb up to the reference mark R. If the atmospheric pressure (in mm of mercury) is *P* then the pressure of the gas is given by

$$p = P + h$$

The temperature of the water bath surrounding bulb A may be varied in the range 0–100°C and pressure measurements taken in this range, keeping the volume constant by returning the mercury level to the mark R in each case. Fig. 11:3(b) shows a typical set of results. The similarity between these results and those shown in fig. 11:2(b) is clear.

Fig. 11:3 Variation of gas pressure with temperature. (a) shows the form of a constant-volume apparatus. (b) Plot of pressure *p* (equal to *P* + *h*, where *P* is atmospheric pressure) against temperature. Graph extrapolates to −273°C.

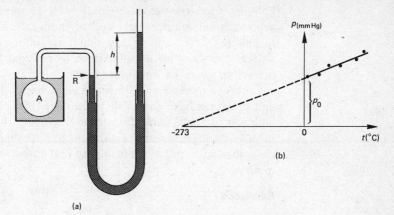

(a)

(b)

It is from experiments such as these that we are able to obtain the absolute scale of temperature discussed in section 10:3. In mathematical terms this is simply a shift of axes, transferring the origin to −273°C, i.e. 0 K.

By combining these empirical laws the **equation of state** for a gas may be obtained:

$$\frac{pV}{T} = \text{constant}$$

This is usually written so that the constant takes account of the mass m of gas being considered and also the nature of the gas. Hence we obtain

$$pV = mR_1T$$

R_1 is usually referred to as the *gas constant*. It has units J/kg K (as for specific heat capacity) and its value varies from gas to gas. If we consider one mole of various gases (e.g. 2×10^{-3} kg of hydrogen, 28×10^{-3} kg of nitrogen, etc.) it is found that the product mR_1 has the same value irrespective of the gas considered, i.e.

$$MR_1 = R$$

where M is the molar mass and R is the **molar gas constant** or **universal gas constant**. Experiment shows that the value of R is 8.31 J/mol K, and **for one mole of any gas**

$$pV = RT \tag{11:4}$$

11:2 Ideal Gases and Real Gases

It is normal to refer to a gas which obeys equation (11:4) as an **ideal gas**. We shall show in section 11:4 that an additional condition for an ideal gas is that the contribution to its internal energy comes from the kinetic energy of the molecules which depends only on the absolute temperature. In practice these are conditions which it is impossible to satisfy and so, strictly, an ideal gas is nothing more than an idea in the minds of theoretical physicists. However it is a very useful idea and one to which the behaviour of **real gases** (such as air, oxygen, nitrogen, hydrogen and helium) approximates reasonably well at moderate temperatures (e.g. around room temperature) and at moderate pressures (around one atmosphere). It is safe to say that any simple laboratory experiment is unlikely to show up the departure of real gas behaviour from ideal gas behaviour given by equation (11:4).

Careful experiments with real gases over a wide range of temperatures and pressures have shown that equation (11:4) has to be modified by the addition of small correction terms involving the pressure and temperature. For our purposes it is sufficient to note that, when we apply the ideal gas equation to problems involving real gases, we are making an approximation.

Example 11:1

A pump initially contains 60 cm^3 of air at 20°C at a pressure of one atmosphere. The gas is then compressed to 15 cm^3 and the temperature rises to 310 K. Determine the final pressure of the air.

Solution The initial temperature is 20°C which is $20 + 273 = 293$ K, and the initial volume is $60 \, \text{cm}^3$ or $60 \times 10^{-6} \, \text{m}^3$.

We are dealing with a fixed mass of gas so that equation (11:4) can be used in the form

$$\frac{pV}{T} = \text{constant}$$

Substituting the initial and final conditions gives

$$\frac{1 \times 60 \times 10^{-6}}{293} = \text{constant} = \frac{p \times 15 \times 10^{-6}}{310}$$

hence $$p = \frac{310 \times 60 \times 10^{-6}}{293 \times 15 \times 10^{-6}} = \frac{310 \times 4}{293} = 4.23 \, \text{atmosphere}$$

Note that, since these equations involve ratios, the conversion of cm^3 to m^3 was not strictly necessary.

11:3 Pressure of an Ideal Gas

In order to use simple kinetic theory (which was introduced in section 10:2) to give an expression for the pressure exerted by an ideal gas, it is necessary to start with a number of basic assumptions. These are:

1 The gas consists of a large number of identical molecules which are in random motion.
2 The molecules occupy a negligible fraction of the total volume available to them.
3 The molecules may be regarded as point masses so that collisions between molecules can be ignored.
4 The molecules experience perfectly elastic collisions with the walls of the container.
5 Any intermolecular forces are negligible and the effect of gravity on the molecules can be ignored.
6 The number of molecules in the sample is large enough for the laws of statistics to be applied.

Fig. 11:4 shows a cube of side l which contains a total number N of molecules, each of mass m, all of which have the same speed c. These molecules are moving around in all directions and, provided that N is large, the average velocity of the molecules will be zero. (This follows since velocity takes account of direction as well as speed and, on average, equal numbers of molecules will be moving in each direction.) We may reasonably consider that the overall picture of the motion is given by $N/3$ molecules moving parallel to Ox; half moving towards ABGF and the other half towards OCDE. Similarly there will be $N/3$ molecules moving in each of the Oy and Oz directions.

A single molecule moving parallel to Ox towards face OCDE will have momentum $-mc$ (since it is moving in the negative Ox direction). When this molecule experiences an elastic collision with face OCDE, its momentum is reversed; the molecule then travels in the positive Ox

direction with momentum $+mc$ and so the change in momentum per collision is

$$mc - (-mc) = 2mc$$

This molecule will travel a total distance $2l$ before it makes its next collision with face OCDE (i.e. it will have bounced off face ABGF). Hence the time between collisions on OCDE is

$$\tau = \frac{2l}{c}$$

and the number of collisions the molecule makes with OCDE per second is

$$f = \frac{1}{\tau} = \frac{c}{2l}$$

Thus the rate of change of momentum per molecule is

$$2mc \times \frac{c}{2l} = \frac{mc^2}{l}$$

Hence for the $N/3$ molecules in the Ox direction the total rate of change of momentum is

$$\frac{N}{3} \times \frac{mc^2}{l} = F$$

This represents the force exerted by the gas on the face OCDE since force is rate of change of momentum (see section 2:3). **Pressure** is defined as **force per unit area**. Hence the pressure on face OCDE is

$$p = \frac{F}{l^2} = \frac{Nmc^2}{3l^3}$$

It is now useful to write N/l^3 as n, which is the number of molecules per unit volume, thus

$$p = \tfrac{1}{3}nmc^2 \tag{11:5}$$

From the definition above, pressure would have units of newtons per metre squared, which is now called the PASCAL (Pa):

$$1\,\text{Pa} = 1\,\text{N/m}^2$$

The pressure has been obtained by considering the rate of change of momentum over unit area of the surface of the container due to gas molecules colliding elastically with the walls. However one weakness of this derivation is that it has assumed that all molecules have the same speed c. This is not the case as experiments with atomic and molecular beams, and more advanced theoretical treatments, both show. At any temperature the molecules possess a range of speeds as shown in fig. 11:5 where $n(c)$ represents the number of molecules possessing speed c. This is known as a Maxwell–Boltzmann distribution.

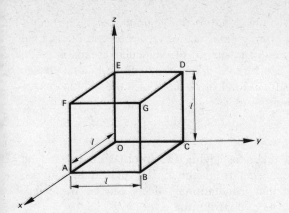

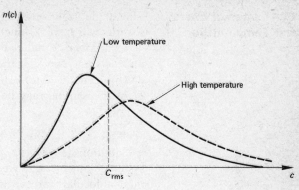

Fig. 11:4 Cube of side l containing gas molecules in random motion.

Fig. 11:5 Spread of molecular speeds at two different temperatures. The root mean square speed at the lower temperature is shown.

When account is taken of this spread of speeds at any temperature, the simple c^2 term that appears in equation (11:5) must be replaced by $\overline{c^2}$, therefore

$$p = \tfrac{1}{3}nm\overline{c^2} \tag{11:6}$$

$\overline{c^2}$ is known as the **mean square speed**. It is calculated by summing the squares of the speeds of all the molecules and dividing the sum by the total number of molecules present. The square root of this quantity is known as the *root mean square* (r.m.s.) speed. The product nm gives the **mass of gas per unit volume**, which is the **density** ρ. Hence the pressure expression becomes

$$p = \tfrac{1}{3}\rho\overline{c^2} \tag{11:7}$$

Example 11:2

Calculate the root mean square speed of oxygen molecules at S.T.P. [S.T.P. is standard temperature and pressure, 0°C (273 K), 1 atmosphere.] The density of oxygen at S.T.P. is 1.43 kg/m³, and the density of mercury (Hg) is 13 600 kg/m³.

Solution **One atmosphere** *in pascals is the pressure exerted by a column of mercury of height* 760 mm. Hence

$$p = dgh$$

where d is the density, g the acceleration of free fall, and h the height.

$$p = 13\,600 \times 9.8 \times 0.76 = 1.013 \times 10^5 \, \text{Pa}$$

From equation (11:7) we can write the mean square speed as

$$\overline{c^2} = \frac{3p}{\rho} = \frac{3 \times 1.013 \times 10^5}{1.43} = 2.125 \times 10^5$$

This quantity is the mean square of the speed and, by taking the square root, we obtain the **root mean square**, or r.m.s., speed which is one measure of the "average" speed of the motion of the molecules. Hence

$$\text{r.m.s. speed } c_{rms} = \sqrt{(\overline{c^2})} = 461 \, \text{m/s}$$

119

11:4 Internal Energy and Temperature

The single molecule that we considered at the beginning of the previous section will have a kinetic energy $\frac{1}{2}mc^2$. However when the effect of a large number of molecules and the spread of molecular speeds is taken into account, the best that can be said is that the average kinetic energy per molecule is $\frac{1}{2}m\overline{c^2}$. Thus for a total of N molecules the kinetic energy U of the sample may be written as

$$U = \tfrac{1}{2}Nm\overline{c^2} \qquad (11:8)$$

Equation (11:8) can be taken as giving the internal energy of the system. (Since we have assumed from the outset that there are no intermolecular forces, and that gravitational effects can be neglected, there will be no potential energy contribution.)

The curves shown in fig. 11:5 and the theoretical treatment of them both suggest that the speed distribution of the molecules depends on the absolute temperature. This dependence is expressed by saying that for an ideal gas, such as this simple kinetic theory picture represents, the internal energy is proportional to the absolute (i.e. kelvin) temperature. That is

$$U = \tfrac{1}{2}Nm\overline{c^2} \propto T \qquad (11:9)$$

Thus by combining equations (11:9) and (11:6), and also taking note that for N molecules contained in volume V, then $n = N/V$, the following results are obtained:

$$p = \frac{1}{3}\frac{N}{V}mc^2$$

$$pV = \tfrac{2}{3}U \qquad (11:10)$$

and $\quad pV \propto T$

Clearly for a fixed mass of gas this last equation is consistent with equation (11:4).

11:5 The First Law of Thermodynamics

In this section we shall consider the effects of heating a sample of an ideal gas. Fig. 11:6 shows a cylinder made from perfectly insulating material. The gas sample is contained in a volume V_1. The piston is also made from insulating material and is fitted in such a way that it can either be locked in position (fig. 11:6(a)) or released so that it can move without friction (11:6(b)) and without allowing any of the enclosed gas to escape. This is clearly a very idealised situation.

Energy is supplied to the sample by means of the electric heater H. Suppose that the heater supplies an amount of **energy** Q to the gas. (Because of the perfectly insulating nature of the material of the cylinder and piston there is no need to consider heat exchange with the surroundings.) The effect of the heating process is to increase the internal energy of the gas, i.e. the temperature of the gas rises. If the piston is locked so that the volume remains constant, the pressure will also rise. If, however,

Fig. 11:6 Fixed mass of gas confined to a cylindrical volume by a piston. Energy is supplied by heater H. (a) Gas at constant volume; (b) gas expands against frictionless piston to keep pressure constant.

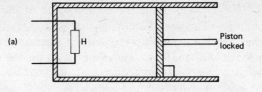

(a)

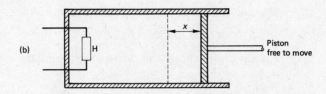

(b)

the piston is not locked, it will move to take up a new equilibrium position such that the pressure on it is the same on both sides. Hence for the situation shown in fig. 11:6(b) it is the pressure that remains constant and the volume increases to some value V_2. The energy Q in this case goes partly to increasing the **internal energy** U of the gas and partly to doing work against the atmosphere by moving the piston. This is usually called **external work**. Thus if the initial internal energy is U_1, corresponding to temperature T_1, and at the end of the heating process the internal energy is U_2 corresponding to temperature T_2, we may write

$$Q = (U_2 - U_1) + W \qquad (11:11)$$

where W represents the external work. This is a very important equation. It is the *law of the conservation of energy* expressed in such a way as to involve internal energy and heat, and is known as the **first law of thermodynamics**.

The external work done in case (b) of fig. 11:6 can be calculated without difficulty. Suppose that the piston moves through a distance x in order to keep the gas pressure constant. If the cross-sectional area of the piston is A, then the force exerted on the piston is pA and the work done by the gas in displacing the piston through distance x is

$$W = pAx$$

But Ax is just the increase in volume, i.e.

$$\Delta V = V_2 - V_1 = Ax$$

$(U_2 - U_1)$, which is the change in internal energy, is frequently written as ΔU. Thus combining the above equations we get

$$Q = \Delta U + p\,\Delta V \qquad (11:12)$$

Case (a) in fig. 11:6 is simply a special case in which $\Delta V = 0$ and no external work is done. It should be pointed out that the first law of thermodynamics has a wide range of applications and is not solely restricted to the case of gases as discussed here.

Example 11:3

0.5 mole of nitrogen gas is confined in a cylindrical volume by a piston of $0.06 \, m^2$ cross-sectional area. $10^3 \, J$ of energy is supplied to the gas and the gas expands moving the piston 0.1 m. Assuming no heat is lost to the surroundings calculate the increase in the internal energy and the increase in the mean KE per molecule. (Avogadro's number $= 6.02 \times 10^{23}$ per mole, atmospheric pressure $= 1.013 \times 10^5 \, Pa$.)

Solution The increase in gas volume is

$$\Delta V = Ax = 0.1 \times 0.06 = 6 \times 10^{-3} \, m^3$$

The gas expands against the atmosphere, hence the external work done by the gas is

$$W = p \, \Delta V = 1.013 \times 10^5 \times 6 \times 10^{-3} = 607.8 \, J$$

Rearranging the first law equation (11:12) gives

$$\Delta U = Q - p \, \Delta V$$

Hence increase in internal energy is

$$\Delta U = 10^3 - 607.8 = 392.2 \, J$$

1 mole of gas contains 6.02×10^{23} molecules, hence this sample of nitrogen contains 3.01×10^{23} molecules. The increase in mean kinetic energy per molecule is therefore

$$\frac{392.2}{3.01 \times 10^{23}} = 1.3 \times 10^{-21} \, J$$

11:6 Molar Heat Capacity of an Ideal Gas

In the previous chapter heat capacity was defined as

$$\frac{Q}{T_2 - T_1} \quad \text{or} \quad \frac{Q}{\Delta T} \quad \text{where} \quad \Delta T = T_2 - T_1$$

From the first law, the heat capacity can be calculated directly, since substituting for Q from equation (11:12) gives

$$\frac{Q}{\Delta T} = \frac{\Delta U + p \, \Delta V}{\Delta T} \tag{11:13}$$

Two particular heat capacities can be identified from this equation. If the volume remains constant, $\Delta V = 0$ and a smaller amount of heat Q is needed to produce a change in internal energy ΔU corresponding to the temperature change ΔT. If we consider one mole of gas, the **molar heat capacity at constant volume**, usually written as C_v is

$$C_v = \frac{\Delta U}{\Delta T} \quad \text{or} \quad \Delta U = C_v \, \Delta T$$

If we use calculus notation, then in the limit as $\Delta T \rightarrow 0$ we have

$$C_v = \frac{dU}{dT} \tag{11:14}$$

Equation (11:10) shows that $U = \frac{3}{2}pV$, and for 1 mole of an ideal gas $pV = RT$ (equation 11:4), hence

$$U = \tfrac{3}{2}RT \quad \text{and} \quad C_v = \tfrac{3}{2}R$$

which is the molar heat capacity of an ideal gas at constant volume.

For the heating process at constant pressure, equation (11:13) gives

$$C_p = \frac{\Delta U}{\Delta T} + \frac{p\,\Delta V}{\Delta T}$$

Again using calculus notation, for $\Delta T \to 0$

$$C_p = \frac{dU}{dT} + p\frac{dV}{dT}$$

i.e. $\quad C_p = C_v + p\dfrac{dV}{dT}$

Differentiating equation (11:4) for one mole of gas under constant pressure conditions gives

$$p\frac{dV}{dT} = R$$

therefore $\quad C_p = C_v + R$ \hfill (11:15)

C_p is the **molar heat capacity at constant pressure**, and using the value for C_v of $\frac{3}{2}R$ gives

$$C_p = \tfrac{5}{2}R$$

The ratio of these heat capacities

$$\gamma = C_p/C_v$$

would therefore be expected to have a value of 5/3. These values apply to a *monatomic* gas.

In a *diatomic* gas, the molecule can have rotational as well as translational energy and the expressions for C_v, C_p and γ are different. (For air, γ has a value of $\frac{7}{5}$ or 1.4.) Equation (11:15) still applies however.

Exercises

For all problems associated with this chapter involving standard pressure and temperature, the following values should be used:

1 atmosphere $= 1.013 \times 10^5\,\text{Pa} \qquad 0°\text{C} = 273\,\text{K}$

11.1 Given that under conditions of standard temperature and pressure, 1 mole of an ideal gas occupies a volume of $22.4 \times 10^{-3}\,\text{m}^3$, obtain a value for the molar gas constant.

11.2 Given that the density of helium at standard temperature and pressure is $0.178\,\text{kg/m}^3$ obtain a value for the unit mass gas constant assuming that helium may be regarded as an ideal gas. The molar mass for helium is $4 \times 10^{-3}\,\text{kg/mol}$; state whether your result is consistent with that obtained for 11.1.

11.3 Using an experimental arrangement as shown in fig. 11:1, a student obtained the following results:

h (mmHg)	l (mm)
383	103
337	107
291	111
246	116
201	121
175	125
115	133
70	140
27	147
−14	156
−55	165
−95	175

Display these results graphically. Do you consider that they are consistent with equation (11:1)? Use your graph to determine the value of the atmospheric pressure at the time that the experiment was performed.

11.4 Using a constant–volume air thermometer of the type shown in fig. 11:3 on a day when the ambient pressure was 751 mm of Hg, a student obtained the following results:

t (°C)	h (mmHg)	t (°C)	h (mmHg)
19.5	37	70.0	169
31.0	69	89.7	220
39.0	88	99.4	242
44.0	104		
55.5	132		

Display these results graphically. Do you consider that they are consistent with equation (11:3)? Use the graph to obtain a value for β; give the value of the intercept on the temperature axis.

11.5 In an experiment, 1.2×10^{-3} m^3 of gas were collected at a pressure of 10^5 Pa and temperature of 20°C. Determine the volume if the gas had been collected under conditions of standard pressure and temperature.

11.6 A cylinder of volume 20 litres contains compressed air at a pressure of 9.8 atmospheres and is located in cold store where the temperature is 6°C. Assuming that air may be regarded as an ideal gas, determine how many moles of air are contained. The cylinder is fitted with a safety valve which will allow gas to escape when the pressure rises above 10 atmospheres. Calculate how much gas will escape when the cylinder is brought into a laboratory where the temperature is 22°C. ($R = 8.31$ J/mol K.)

11.7 A vessel of volume 60 litres contains air, which may be assumed to be an ideal gas, at atmospheric pressure (760 mmHg). It is connected to an exhaust pump, the cylinder of which has a maximum volume of 2 litres. Assuming that the temperature remains constant throughout, calculate the pressure in the vessel after a) 1 cycle, b) 2 cycles of the pump. Determine how many cycles of the pump would be needed to reduce the pressure below 1 mmHg, assuming that there was no leakage of air into the system.

11.8 Two vessels, A and B, are connected by a tube whose volume may be neglected, and the system contains a gas which may be regarded as ideal. A has a volume of 1.3 litres and B has a volume of 2.8 litres. In a certain process, A is maintained at 0°C while B is heated from 21°C to 119°C. Obtain a value for the pressure in the system at the end of the process, given that the initial pressure was 750 mmHg.

11.9 Calculate the root mean square speed for nitrogen molecules under conditions of standard temperature and pressure. Nitrogen has a relative molecular mass of 28 and you may assume that the molar volume at S.T.P. is 22.4×10^{-3} m^3.

11.10 Using the data given in 11.9 estimate the root mean square speed for nitrogen molecules at 373 K, assuming that the pressure remains atmospheric.

11.11 One mole of argon, which may be regarded as an ideal gas, is contained in a cylinder of cross-sectional area 0.05 m^2. The sample is enclosed by a piston which is capable of moving freely. The external pressure is atmospheric. In an expansion process, 300 J of energy were supplied to the gas and the piston moved a distance of 2.4×10^{-2} m. Calculate the increase in the internal energy of the sample. Given that the molar heat capacity of argon at constant volume is 12.5 J/mol K, obtain a value for the rise in temperature of the sample.

11.12 Answer all *three* parts
a) Explain what is meant by the term: *ideal gas*. Describe an experiment to investigate the way in which the volume of a fixed mass of an ideal gas varies with pressure under conditions of constant temperature. Sketch the results that are likely to be obtained from such an experiment.
b) A sample of a certain gas which may be regarded as ideal occupies a volume of 1 litre under a pressure of 15 atmospheres at 20°C. The conditions are then changed to a volume of 10 litres and a pressure of 1 atmosphere. Determine the temperature under these conditions.
c) The kinetic model of an ideal gas sets out to predict the bulk behaviour of the gas in terms of the motion of individual molecules. State the assumptions that have to be made in order to obtain an expression for the pressure of a gas according to this model. (NB. You are not asked to derive the expression.) Explain the significance of the *mean square speed* of the molecules in this context.

12 The Liquid State

In this chapter we shall draw together some of the ideas set out in chapters 10 and 11. We shall seek to take some of these ideas a little further and to distinguish between gases and vapours, before proceeding to consider some of the properties of liquids.

12:1 Vapour Pressure and Saturation

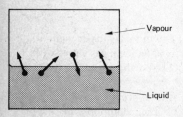

Fig. 12:1 illustrates a closed container which is about half full of liquid at room temperature. The space above the liquid surface is filled with a *mixture of air and vapour*. The number of molecules in the vapour phase† at any moment will depend on the rate at which molecules are escaping from the liquid and also on the rate at which molecules are being recaptured by the liquid from the vapour. When the liquid is first put into the container, the evaporation rate will be greater than the rate of recapture, so the vapour will grow at the expense of the liquid. We can think of these vapour molecules exerting a pressure, according to simple kinetic theory, in their own right; this is the **vapour pressure**. As the number of molecules in the vapour phase increases, so the recapture rate will also increase. Eventually the rate of recapture will become equal to the rate of escape and a state of **dynamic equilibrium** will exist between the vapour and the liquid at the prevailing temperature. The evaporation and recapture processes are still going on but, as the rates are now equal, the numbers of molecules in the respective liquid and vapour phases will remain constant. Under these conditions the vapour is said to be **saturated**. The molecules in the vapour phase will exert a pressure, which can be determined by the kinetic model of the previous chapter. This pressure is known as the **saturated vapour pressure** (s.v.p.). For water at room temperature (20 °C) the value of the saturated vapour pressure is about 18 mm of mercury.

If the temperature of the sample as depicted in fig. 12:1 is increased, the rate of evaporation will increase and a new dynamic equilibrium will be established at this higher rate, with more molecules in the vapour phase. As the kinetic theory expression for pressure depends on both the number of molecules per unit volume and the mean square speed ($p = \frac{1}{3}nm\overline{c^2}$), we see that there are two factors acting to increase the saturated vapour pressure. Hence we would expect the s.v.p. to rise

Fig. 12:1 Diagrammatic representation of liquid and vapour coming to equilibrium in an enclosed container.

† We are using the word "phase" here in a rather precise manner. A phase is a region within a sample where its physical properties, e.g. density, refractive index, specific heat capacity, etc., are all well defined. The word "state" is often used loosely with this meaning, as in the title of this chapter! Strictly, the word "state" refers to the values of pressure, volume and temperature which describe the condition of the sample at any instant.

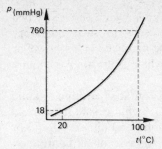

Fig. 12:2 S.V.P./temperature curve for water.

rapidly, especially at higher temperatures. The variation of the s.v.p. of water with temperature is shown in fig. 12:2.

In chapter 10, we took the boiling point of pure water at 1 atmosphere pressure to be exactly 100 °C: the upper fixed point. Fig. 12:2 shows that at 100 °C the value of the s.v.p. is 760 mmHg, which is 1 atmosphere $(1.013 \times 10^5 \, Pa)$. Hence it can be stated that **boiling** takes place when the s.v.p. is equal to the ambient pressure. This fact is made use of in the domestic pressure cooker. By placing "weights" over a vent in the lid, an excess pressure can be built up inside the cooker. For an excess pressure of 1 atm. (15 lb. wt), cooking can take place at a temperature of about 120 °C; this leads to a significant reduction in the total cooking time and a corresponding saving in fuel.

Boiling differs from evaporation in that whereas evaporation can, and does, take place at any temperature below the boiling point, and occurs only from the free surface of the liquid, boiling occurs at a well-defined temperature depending on the pressure, and takes place in the bulk of the liquid. When the liquid boils, small pockets of vapour are formed on any convenient nuclei present in the body of the liquid; these pockets of vapour rise to the surface, expanding as they do so, and break as bubbles giving the familiar agitated, seething appearance of a boiling liquid. When very pure liquids are being heated in clean containers, suitable nuclei for bubble formation may not be present; in this case the liquid **superheats**: its temperature rises above the normal boiling point. When boiling does commence it does so explosively and hot liquid may be shot out violently—this process is known as *bumping*. It is for this reason that students doing laboratory work are instructed not to point boiling tubes towards themselves or other people in the laboratory. Bumping can be prevented by introducing suitable nuclei, usually small pieces of some inert substance such as glass beads or fragments of broken porcelain into the container. These "anti-bumping stones" assist bubble formation at the normal boiling point.

12:2 Two Historic Experiments

We shall now refer briefly to two experiments performed over a hundred years ago but which throw light on the question of the liquefaction of gases.

Around 1822, Caignard de la Tour showed that alcohol was able to remain liquid at temperatures above its normal boiling point provided that the pressure was increased. He also found that there was a *critical temperature*, above which it was not possible for the alcohol to exist as liquid. Subsequent studies by Andrews and others showed that all gases have a critical temperature above which they cannot be liquefied merely by increasing the pressure.†

It is usual to restrict the use of the term *vapour* to a substance in its gaseous form below its critical temperature.

† To follow up the fascinating story of the liquefaction of gases, the production of low and very low temperatures, and the properties of materials at these temperatures, consult the excellent little book by Kurt Mendelssohn entitled *The Quest for Absolute Zero* (World University Library, 1966).

12:3 Surface Tension and Angle of Contact

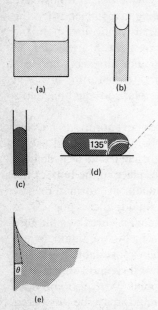

Fig. 12:3 (a) Beaker containing water, (b) Narrow glass tube containing water, (c) Narrow glass tube containing mercury, (d) Mercury globule resting on a clean glass plate, (e) Angle of contact effectively zero for a liquid which wets the solid.

A **surface**, or an interface, is the boundary region which separates two phases. The structure and properties of materials in such boundary regions are of considerable interest; they may be highly complex however and Surface Science is an active research area which spills out over many other disciplines.

Probably the first surface that comes to mind is that separating a liquid from air. It is worth noting however that much of what is said here could apply equally well to the boundary between a solid and either a liquid or air, or that between two immiscible liquids, e.g. a layer of oil floating on water. Fig. 12:3 shows a variety of liquid/air surfaces. These diagrams show familiar situations where water and similar liquids tend to creep up the sides of the vessel containing them giving a large flat surface (fig. 12:3(a)) or a nearly hemispherical surface (fig. 12:3(b)), whereas mercury in contact with glass shows a surface which is curved in the opposite direction (fig. 12:3(c)). The case of a mercury globule resting on a horizontal, clean glass plate is particularly interesting; fig. 12:3(d) represents a cross-section through the centre of the globule. Provided that the globule is large, say 25 mm or more in diameter, the upper surface will be horizontal and the sides will be curved as indicated, making contact with the glass plate at an angle of 135°. This is known as the **angle of contact**. In the case of water and a clean glass surface, the angle of contact is usually taken to be zero; the water in the beaker and tube of figs. 12:3(a) and (b) can be thought of as creeping up the wall of the container to give a value of θ very close to zero as indicated in fig. 12:3(e).

It is usual to explain the shapes of surfaces and the existence of an angle of contact in terms of the forces between molecules. Such forces between molecules of the same kind (e.g. two water molecules) are known as **cohesive**, whereas forces between different kinds of molecule (e.g. water molecules and the molecules in the glass wall of the container) are known as **adhesive**. If adhesive forces are stronger than cohesive forces then the liquid creeps up the wall of the container or spreads over a horizontal surface; in the opposite case, where the cohesive forces are stronger than the adhesive, a liquid such as mercury experiences a downward curvature at the container walls or forms a stationary droplet on a horizontal surface (see fig. 12:4).

The existence of a curved surface implies that it is stretched and is, therefore, in a state of tension; in fact any surface, whether curved or plane, exhibits this **surface tension**. We can use the same ideas about intermolecular forces that were helpful in discussing the *meniscus* effect in narrow tubes, etc. (fig. 12:3) to aid the understanding of surface tension. The liquid is held together by attractive forces between the molecules and there is an optimum separation between molecules. Fig. 12:5 shows in general terms how the intermolecular force between two molecules varies with the separation between their centres. At the surface, where a transition from liquid to air/vapour is occurring, the intermolecular spacing is increasing rapidly and there is an accompanying decrease in the density. This can perhaps be shown up by reference to fig. 12:6. It is helpful to think of a *sphere of influence* around each molecule, of radius

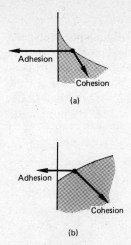

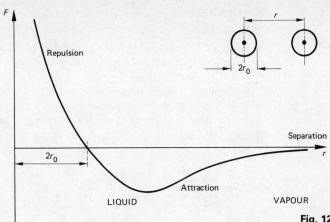

Fig. 12:4 Adhesive and cohesive forces in cases where the liquid (a) does, and (b) does not, wet the solid.

Fig. 12:5 Intermolecular force between two molecules as a function of the separation between their centres.

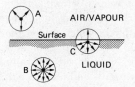

Fig. 12:6 Imbalance of intermolecular forces for molecule in the surface of a liquid.

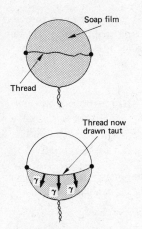

Fig. 12:7 Surface tension forces acting on a light thread in a soap film when one portion of the film is removed.

equal to several molecular diameters such that it extends to a point where the force in fig. 12:5 has a small value.

A molecule such as A, indicated in fig. 12:6, has only a few other molecules coming within its sphere of influence and such intermolecular forces that it experiences will in general be very weakly attractive. A molecule such as B, deep in the body of the liquid, has many more molecules within its sphere of influence; the forces that it experiences will in general be attractive but distributed fairly uniformly around it so that the resultant force on B remains small. However, it is the molecule such as C which is of particular interest; here the sphere of influence is partly within the liquid where the density of molecules is high and partly in the vapour/air where the density of molecules is very much lower. Thus, these surface molecules as typified by C experience a marked imbalance in the forces acting on them. The component of these forces perpendicular to the surface tends to draw surface molecules into the body of the liquid, while the horizontal components give rise to a state of tension in the surface.

The surface tension is defined in terms of the force acting in the surface and at right angles to a line of unit length in the surface. *Surface tension forces act in such a way as to minimise the surface area.* These points may be illustrated by considering the case of a soap-water film contained in a wire loop. If a light thread is attached loosely across the diameter of the loop, as shown in fig. 12:7, and one portion of the film is then pricked with a needle, the remaining portion of the film will take up the smallest possible surface area, and the thread will be drawn tight by surface tension forces acting as shown.

The usual symbol to denote surface tension is γ and it is clear from the definition that the units will be newtons/metre (N/m). At room temperature the surface tension of pure water in contact with air saturated with water vapour is about 72×10^{-3} N/m.

129

Example 12:1

A clean glass microscope slide of length 75 mm, width 25 mm and thickness 3 mm is suspended vertically with its long edge horizontal from one arm of a beam balance and is counterpoised (as shown in fig. 12:8). A large beaker of clean water of surface tension 72×10^{-3} N/m is then carefully brought up on a rising table until the slide is immersed to a depth of 12 mm. Calculate the change that must be made to the counterpoising mass if the balance is to be restored.

Assume that the density of water is 1000 kg/m^3 and that the acceleration of free fall is $g = 9.8$ m/s^2.

Fig. 12:8

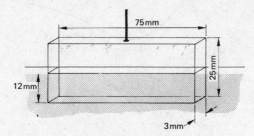

Solution   The forces acting upon the slide are (i) the weight W which acts vertically downwards, (ii) the upthrust U which is equal to the weight of the volume of water displaced, and (iii) the surface tension forces which act on the line of contact with the slide and the water. Provided that the angle of contact is zero—we may assume that this is the case since the slide is clean and the water pure—these surface tension forces will act vertically downwards and tend to pull the slide down into the water.

Hence, we may write for the total force acting on the slide:

$$F = W - U + \gamma l$$

where γ denotes the surface tension and l is the length of the line of contact between the slide and the water.

Since the slide is initially counterpoised, the weight W need not enter directly into the calculation and the change in the counterpoising mass, which we can write as m, will reflect the effect of the upthrust and surface tension forces. Thus, when balanced,

$$mg = \gamma l - U$$
$$= [72 \times 10^{-3}(75 + 3 + 75 + 3)10^{-3}] - [(75 \times 3 \times 12)10^{-9} \times 1000 \times 9.8]$$
$$= [1.12 \times 10^{-2}] - [2.65 \times 10^{-2}]$$
$$9.8m = -1.53 \times 10^{-2} \qquad m = -1.56 \times 10^{-3} \text{ kg}$$

Hence, in order to restore the balance, the counterpoising mass must be reduced by 1.56 g. This means that the surface tension forces are not sufficient to balance the upthrust at this depth of immersion.

The discussion of surface tension in terms of the molecular theory indicates that the processes which govern evaporation [attractive forces which act to prevent molecules escaping from the liquid surface so that only the fastest molecules are able to get away as mentioned in sections

10:2 and 10:8] and surface tension [attractive forces seeking to minimise the surface area] are basically very similar. It can be shown both by experiment and by theoretical treatment that the surface tension and the latent heat of evaporation for a liquid both decrease with temperature. Well away from the critical temperature, as in the case of water between 0°C and 100°C, this decrease is linear to a good approximation; at temperatures close to the critical value the surface tension and latent heat of evaporation both decrease very rapidly indeed and become zero at the critical temperature.

12:4 Capillarity

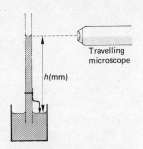

Fig. 12:9 Experimental measurement of capillary rise.

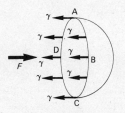

Fig. 12:10 Schematic representation for the surface tension forces acting on a hemispherical surface in equilibrium, giving rise to the excess pressure on the concave side.

If a piece of clean glass capillary tube is supported vertically, with its lower end dipping into a beaker of water, or a similar liquid, it can be observed that the liquid will rise up the tube above the level of the free surface of the liquid outside. This phenomenon is known as **capillarity** or *capillary rise*. The height to which the liquid rises depends on the bore of the tube, and it can easily be demonstrated by using tubes of different bores that the finer the bore the higher the rise will be. A close inspection of the meniscus would reveal that it is of hemispherical shape as shown in fig. 12:3(b). In order to measure the rise, a travelling microscope may be focused through the wall of the tube on to the meniscus; a small metal probe may be used to locate the free surface of the liquid in the beaker. Such an experimental set-up is shown in fig. 12:9.

In order to explain the phenomenon of capillary rise it is necessary to recognise that a pressure difference exists across a curved surface, with the higher pressure on the concave side. A helpful analogy here may be obtained by thinking of a sheet pegged out on a washing-line on a windy day. The wind blowing into the sheet causes it to belly out on the low-pressure side and as the pressure increases [i.e. the wind gets stronger] on the high-pressure side so the curvature of the sheet becomes more pronounced, i.e. the radius of curvature becomes smaller.

Consider a hemispherical surface of radius r as drawn in fig. 12:10. ABCD is the open circular edge of the surface. Surface tension forces γ per unit length will act around ABCD in a direction perpendicular to the plane of ABCD. The length of ABCD is $2\pi r$; therefore the total surface tension force acting towards the left is $2\pi r\gamma$. However, since the surface is in equilibrium there must be an equal and opposite force F acting into the concave side of the surface such that

$$F - 2\pi r\gamma = 0$$

It is convenient to regard F as the product of an excess pressure Δp and the area ABCD, i.e. $F = \pi r^2 \Delta p$.

$$\pi r^2 \Delta p - 2\pi r\gamma = 0$$

$$\therefore \quad \Delta p = \frac{2\gamma}{r} \qquad (12:1)$$

We can now use equation (12:1) to obtain an expression for the height to which a liquid will rise up a capillary tube. We shall suppose that we are

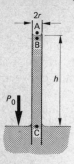

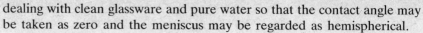

Fig. 12.11 Application of excess pressure to determine the height of capillary rise.

dealing with clean glassware and pure water so that the contact angle may be taken as zero and the meniscus may be regarded as hemispherical.

In fig. 12:11, A is a point just outside the meniscus; the pressure at this point is atmospheric, P_0. Therefore, according to equation (12:1), the pressure at B, just inside the meniscus, will be

$$p_B = P_0 - \frac{2\gamma}{r}$$

Since the column of liquid is standing in equilibrium, the pressure p_C at the point C inside the tube but in the same horizontal line as the free surface of the liquid outside must be equal to the atmospheric value P_0. For the liquid in the tube we may write that the pressure at C is given by the value of p_B plus the hydrostatic pressure $g\rho h$ due to the column of height h and density ρ. Thus for the pressure at C we have:

$$P_0 = P_0 - \frac{2\gamma}{r} + g\rho h$$

which gives $\quad h = \dfrac{2\gamma}{g\rho r}$ \hfill (12:2)

Example 12:2

Calculate the height to which water would rise up a clean glass capillary tube of radius 0.3 mm given that the surface tension of water is 72×10^{-3} N/m, the density of water is 1000 kg/m^3 and assuming a value of 9.8 m/s^2 for the acceleration of free fall.

Solution By direct substitution of the values into eqn. (12.2) we obtain:

$$h = \frac{2 \times 72 \times 10^{-3}}{9.8 \times 1000 \times 0.3 \times 10^{-3}} = 49 \times 10^{-3}\,\text{m} = 49\,\text{mm}$$

So far we have assumed that water, or the liquid considered, wets the glass perfectly; this is simply another way of saying that the angle of contact is zero. When this is not the case the meniscus is no longer hemispherical (see fig. 12:12). The vertical component of the surface tension contributing to the excess pressure is now $\gamma \cos \theta$, so equation (12.2) modifies to:

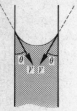

Fig. 12:12 Effect of non-zero angle of contact in capillary rise.

$$h = \frac{2\gamma \cos \theta}{g\rho r}$$ \hfill (12:3)

Equation (12:1) applies to the excess pressure across a single hemispherical surface separating a liquid from air. A soap bubble, such as that shown in fig. 12:13, is a very thin liquid film in the form of a hollow sphere. The internal pressure p_i exceeds that at point A inside the bubble wall by an amount $2\gamma/r$. Furthermore, if the wall is very thin, the pressure at A exceeds the external pressure p_0 by a similar amount. Hence, by addition, the total excess pressure is

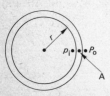

Fig. 12:13 Excess pressure inside a soap bubble in air.

$$\Delta p = p_i - p_o = \frac{4\gamma}{r}$$

12:5 Surfactants, Contaminants and Detergents

In the most general terms it is usual to regard a **surfactant** as being any substance which changes the surface properties, particularly the surface tension and contact angle, of a sample. The effects of surfactants are most clearly observed in terms of the spreading of liquids over solids.

In the preceding sections we have been at pains to point out that we have been considering clean surfaces and pure liquids, i.e. cases where surfactants are absent. The commonest surface **contaminant** is grease. The human skin contains natural oils and grease which are essential for its biological function but which can have unfortunate consequences for a "clean experiment". A single fingerprint on a piece of clean glassware will significantly affect the way in which it will be wetted by a liquid; thus in such experiments, tweezers, remote-handling tools, etc. will be used and the experimenter will wear sterile gloves. On a more lowly level, anyone who has attempted to wash up greasy plates will be familiar with the difficulty of getting the water to wet the plates. The grease layer acts as a contaminant which reduces the value of the surface tension of the solid to one which is comparable with that of the water; hence the water forms globules which stand on the contaminated surface. Hot water may help here, as it will melt the grease and reduce the surface tension of the water slightly. However, the usual and most effective way to deal with this problem is to add a small quantity of a proprietary brand of **detergent** to the water. The combination of hot water and a surfacting detergent significantly reduces the surface tension and contact angle of the water and allows wetting to take place. From a chemist's point of view, the action of a detergent may be thought of as an attack on the grease, surrounding it with a monomolecular layer which has a hydrophilic (water-seeking) outer surface.

A similar problem involving wetting is encountered in the process of soldering, where the first prerequisite is to degrease and clean up the metal surfaces to be joined. Liquid metals in general have high surface tension values; this makes it rather difficult to get the molten solder to flow. To overcome this tendency for the solder to stand as a globule, which will lead to a "dry joint", a flux is used with the solder to reduce its surface tension and contact angle.

12:6 Viscosity

There are many ways in which a one-sentence definition of a liquid may be attempted; none of which are likely to be complete or entirely satisfactory. For the present purpose, we wish to make use of the statement that: *a liquid is a substance which is unable to sustain a shear stress.* Fig. 12:14 shows a sample of a liquid trapped between two flat parallel plates; the upper plate can be moved horizontally relative to the fixed lower plate. Thus, the liquid is acting essentially as a lubricant.

Suppose that a horizontal force F_x is applied to the upper plate and that the area of contact between the liquid and the plate is A; thus the shear stress on the sample is F_x/A. However, the liquid will exert a resistance or *viscous drag* on the motion of the plate. Clearly, if the plate is moving with a constant velocity V_x, the viscous drag is equal and

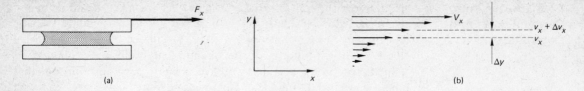

(a)   (b)

Fig. 12:14 (a) Laminar flow for a viscous liquid between two plates; (b) shows the velocity gradient across the liquid.

opposite to F_x. With a solid sample, elastic behaviour would be expected for a small applied stress, whereas in the liquid a situation known as *laminar flow* is set up.

We may think of the molecules in the liquid as being located in horizontal planes or laminas. The lamina in contact with the lower plate is at rest, while that in contact with the upper plate has the velocity V_x. The laminas between these extreme examples are all moving with different velocities, and a velocity gradient $\Delta v_x / \Delta y$ will be set up in the vertical direction as indicated in fig. 12:14(b).

With the usual calculus notation we can write the velocity gradient as

$$\frac{dv_x}{dy} \text{ in the limit as } \Delta y \text{ tends to zero.}$$

The **coefficient of viscosity** η for a fluid relates the applied shear stress to the velocity gradient, so we may write:

$$\frac{F_x}{A} = \eta \frac{dv_x}{dy} \tag{12:4}$$

The units for viscosity can be shown to be kg/m s; the SI name given to this unit is the POISEUILLE. (NB. This should not be confused with an earlier unit known as the poise which belongs to the centimetre-gram-second system. 1 poiseuille = 10 poise.) Around room temperature, water has a viscosity of about 1 milli poiseuille, while a substance like treacle (syrup) might have a value between 100 and 200 poiseuille. The viscosity of a substance gives a measure of how easily it flows; a low value for the viscosity corresponds to easy flow. In general the value for the viscosity of any substance decreases as the temperature is raised.

The behaviour described above is known as Newtonian viscosity. In practice, many substances of technological and commercial interest are non-Newtonian in that their viscosity depends on the rate of shear. Examples of this kind of behaviour can be drawn from the paint and food industries, e.g. non-drip paints have a low viscosity when the rate of shear is high (they flow easily while they are being brushed) but a high viscosity at zero shear (they cease to flow when the brushing stops). Similarly, considerable research effort has been put into the development of the margarines that "spread straight from the fridge".

The example considered above in fig. 12:14 showed the action of a lubricant operating between two moving parts; this has clear applications in mechanical engineering. In the motor industry particularly it is impor-tant to have lubricating oils whose properties remain constant under varying conditions of temperature and shear rates—the visco-static oils.

These are all examples of non-Newtonian viscosity; however, the consideration of Newtonian behaviour provides a good introduction to a topic of considerable interest and a field of active research and development.

Exercises

12.1 A simple barometer was set up and the height of the mercury column was observed to be 757 mm. The air temperature in the laboratory was 21°C. A small quantity of water was carefully introduced at the base of the mercury column; more water was introduced until a thin film of water was observed to be resting on the mercury meniscus at the top of the column. It was then observed that the height of the mercury column was only 738 mm. Use these data to obtain a value for the saturation vapour pressure of water at 21°C; explain the physical principles which underlie the method of obtaining this result.

(NB. Students are advised that this is an experiment which should only be attempted under the strictest supervision.)

12.2 Using the data given in worked example 12.1, determine the depth of immersion for the slide at which no change in the counterpoising mass would be required. Determine also the force needed just to lift the slide clear of the water.

12.3 A drinking straw has a diameter of 3 mm. Estimate the pressure with which someone must blow down it in order to form a hemispherical bubble when the tip is immersed to a depth of 10 cm in a beaker of water. (Density of water = 1000 kg/m^3; g = 9.8 m/s^2; surface tension of water = 72×10^{-3} N/m.)

12.4 Water of density 1000 kg/m^3 and surface tension 72×10^{-3} N/m rises to a height of 48 mm in a given capillary tube. Calculate the height to which alcohol of density 800 kg/m^3 and surface tension 23×10^{-3} N/m would be expected to rise up the same tube. (Assume that the angle of contact for alcohol is zero.)

12.5 Using a set-up similar to that shown in fig. 12:14, it was found that if an oil film 0.1 mm thick was used to lubricate two metal plates of area of contact 25×10^{-4} m^2 a force of 0.6 N applied tangentially to the upper plate would cause it to move with a steady velocity of 2×10^{-2} m/s. Estimate a value for the viscosity of the oil.

13 Electrostatics

A careful study of the behaviour of electric charges can lead us to a complete theory of electricity and magnetism. This theory finds application in devices which affect all our lives, whether it be for example, television, radio, the electric motor or the telephone. It may be appreciated therefore, that a grounding in the principles involved is of great importance to the scientist or engineer. In the following chapter we consider the behaviour of stationary charges, while subsequent chapters will be concerned with charges in motion.

13:1 Charge

A body will accelerate so long as there is a resultant force acting on it (see section 2:3). A force can be applied in the form of a push or pull, i.e. there is contact between the agent supplying the force and the body which we wish to move. In nature, there are several examples of forces that act over a distance, there being *no contact* between the bodies involved. Such an interaction is often termed **action at a distance**. An example of this is the force of gravity (see section 2:3). The force of gravity is responsible for planetary motion and therefore must act over vast distances of empty space. We now consider another example of action at a distance.

If a rod made from ebonite is rubbed briskly with a piece of fur, and then is held close to some small pieces of paper, the pieces of paper "jump" towards the rod. It cannot be the force of gravity that is responsible for the force experienced by the paper—if it was, then why isn't the paper attracted to the rod before the rod is rubbed? We consider that a **charge** has been built up on the rod and that there is an *electrostatic force of attraction* between the rod and the pieces of paper. An electrostatic force is the force between *stationary* charges.

In nature, there are two types of charge which are termed **positive** and **negative**. If two bodies are charged positively each will experience a force of repulsion when brought close to the other (see fig. 13:1(a)). Similarly two bodies charged negatively will each experience a force of repulsion when brought into close proximity (see fig. 13:1(b)). However two bodies, one charged positively and the other negatively, will experience a force of attraction (see fig. 13:1(c)). This can be summarised as follows:

Like charges repel. Unlike charges attract.

Note that some bodies can be charged directly by friction, others by contact with a body that has been charged previously by friction.

Coulomb's Law It can be shown experimentally that the force between charged bodies decreases with increasing separation. A French physicist, Coulomb, showed that the force between charged spheres obeys an "inverse square law", i.e. if the force between the charged spheres is F and their separation is r, then:

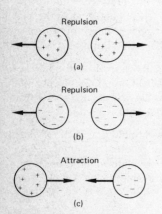

Fig. 13:1 Summary of direction of forces acting on charged bodies. (a) Both bodies carry positive charge; (b) both bodies carry negative charge; (c) one body carries positive charge the other carries negative charge.

$$F = C \times \frac{1}{r^2} \qquad C \text{ is a constant}$$

So, for example, if the distance between the charges is doubled, i.e. it is now $2r$, the force between the charges becomes

$$F' = C \frac{1}{(2r)^2} = \frac{C}{4r^2} = \frac{F}{4}$$

Doubling the distance between the two charged bodies has resulted in a decrease of the force to a quarter of its original value.

The value of the force between charged bodies also depends on the magnitude of the charge on each body. For two metal spheres, the centres of which are a distance r apart, carrying charges Q_1 and Q_2 respectively, the force between them is given by

$$F = k \frac{Q_1 Q_2}{r^2} \qquad (13:1)$$

k is a constant which has the value $\dfrac{1}{4\pi\epsilon_0}$ where ϵ_0 is the **permittivity** of free space.

The force between two charges also depends upon the medium in which they are situated. ϵ_0 is the factor which modifies the constant k for charges placed in free space (i.e. a vacuum). ϵ_0 has the value 8.85×10^{-12} farads per metre. (The Farad is defined in section 13:9.) In practice we shall ignore the very slight modification to k when the charges are placed in air and treat this as a vacuum situation. Coulomb's law can therefore be written fully as

$$F = \frac{Q_1 Q_2}{4\pi\epsilon_0 r^2} \qquad (13:2)$$

In the SI system of units the unit of charge is the COULOMB (symbol C). The coulomb is derived from the ampere which is defined in section 15:9. Equation (13:2) can be used for point charges (those are charges that have negligible dimensions) or uniformly charged spheres. In the latter case we can consider that the charge on the surface of the sphere is concentrated at the centre of the sphere and therefore acts like a point charge.

Example 13:1

Two metal spheres are held with their surfaces 10 cm apart. Each sphere has a radius of 1 cm and carries a positive charge of 2×10^{-4} C. Draw a diagram showing the direction of the electrostatic force acting on each sphere. Calculate the size of the force on each sphere. If each sphere has mass 2 g calculate the gravitational force acting on each mass. Give the ratio of the electrostatic force to the gravitational force.

Solution   The charge on each sphere acts as though it were at the centre of the sphere; therefore the distance between the charges is taken to be 12 cm and not 10 cm. Referring to fig. 13:2 and using equation (13:2),

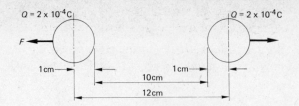

$$F = \frac{Q_1 Q_2}{4\pi\epsilon_0 r^2} = \frac{2\times10^{-4}\times2\times10^{-4}}{4\pi\times8.85\times10^{-12}\times(12\times10^{-2})^2}$$

$$= 25\,000\,\text{N} \quad \text{(rounding up)}$$

In order to calculate the gravitational force between the spheres we must use the equation given in section 2:3 (p. 9):

$$F = \frac{Gm_A m_B}{r^2} = \frac{6.67\times10^{-11}\times2\times10^{-3}\times2\times10^{-3}}{(12\times10^{-2})^2}$$

$$= 1.9\times10^{-14}\,\text{N} \quad \text{(rounding up)}$$

The ratio of the electrostatic force to the gravitational force is therefore 1.3×10^{18}. It can be seen that the gravitational attraction is negligible compared to the electrostatic force of repulsion.

13:2 The Atomic Nature of Charge

All matter consists of atoms (see section 9:1). The atom itself is divisible into three types of particle: protons, neutrons, and electrons. Fig. 20:1 (p. 214) shows a diagram of a typical atom. The proton and the electron carry charge while the neutron does not. We identify negative charge as being carried by the electron and positive charge as being carried by the proton. It was stated that an ebonite rod could be charged negatively by rubbing it with a piece of fur. Prior to rubbing, the rod consisted of equal numbers of protons and electrons. In order for the rod to become negatively charged, electrons have been transferred from the fur to the rod, leaving the fur with a net positive charge and the rod with a net negative charge. When a body is charged positively, we infer that it is deficient in electrons and, when a body is charged negatively, it has an excess of electrons.

Metals in general are good conductors of charge. A good **conductor** is a material which has a significant number of electrons that are loosely bound to their atoms (often referred to as "free" electrons). In copper, for example, there are many electrons that can move quite freely within the copper and can distribute themselves very quickly when they are subjected to an electrostatic force. If you hold a copper rod in your hand, then attempt to transfer some electrons onto it by contact with a charged ebonite rod, you would find that the copper rod would conduct the electrons from the point of contact of the two rods to your hand immediately, maintaining the electrical neutrality of the copper.

When a glass rod is charged by rubbing, it is found that the charge does not distribute itself throughout the glass. Glass is termed a good **insulator**. All the electrons surrounding the atoms in the glass are firmly bound to those atoms and are unable to move, i.e. the glass possesses no free electrons. A brief list of conductors and insulators is given in Table 13:1.

Table 13:1

INSULATOR	CONDUCTOR
Glass	Copper
Ebonite	Aluminium
Mica	Silver
PTFE	Sodium
Perspex	Iron
Rubber	Magnesium
Polystyrene	Graphite

13:3 Practical Problems Caused by Charged Bodies

The human body is very sensitive to a flow of charge. Touching a body that is carrying a large charge can give you an electric-shock. The movement of relatively small amounts of charge through the body can be lethal. High voltages can cause such a movement of charge and therefore the human body must be insulated from these voltages. For this reason the casing of electric plugs and sockets is made from rubber or plastic which are good insulators. Electric cables are surrounded by a good insulator such as PVC for the same reason. Under no circumstances should any protective insulating material be removed from a piece of electrical equipment that operates at high voltages, for this could render it deadly.

When handling electronic components, it is often important to ensure that no charge can be built up on them, otherwise the component can become irreparably damaged. Such charge-sensitive components include integrated circuit chips. Elaborate precautions are taken to ensure that charge build-up cannot occur. For example, the components can be pressed into conducting foam which is earthed (i.e. there is a conducting path from the foam to the earth) so that any charge accidentally transferred to the foam is instantly conducted to earth.

13:4 The Electric Field

There is another way of looking at the forces between charges. We may imagine that there is something in the space surrounding a charge that will exert a force on another charge placed in that space. This "something" is called the electric field. An **electric field** is caused by stationary charges and will exert a force on any other charge placed nearby. The magnitude of the force exerted depends on the strength of the electric field and the size of the charge. By knowing the electric field at any point in space, it is possible to determine the magnitude and direction of the force on any charge placed at that point.

The electric field at a point is defined as the force that would be exerted on a unit charge placed at that point in the electric field. If a charge Q

experiences a force F in an electric field, then the electric field strength at the point where the charge is positioned is given by

$$E = \frac{F}{Q} \qquad (13:3)$$

where E is the symbol used to denote electric field. The unit for electric field is the NEWTON PER COULOMB (symbol N/C). The electric field at a point acts in a specific direction; therefore it is a vector quantity. (In a more rigorous treatment of electric fields, the term electric field strength should be used for the vector quantity E which we simply call the electric field.) In principle the electric field at a point in space could be determined by measuring the force on a small charge placed at that point in the field and noting the direction in which the force acts.

A useful way of visualising an electric field is with the aid of **electric field lines**. Fig. 13:3 shows how electric field lines are conventionally drawn. The lines originate from positive charges and terminate on negative charges. The arrow on the field line indicates the direction of the force that a positive charge would experience if placed at that point in the field. The density of field lines (the number of lines per unit area) gives a measure of the strength of the electric field. The greater the density of the field lines, say close to a charge, the greater the magnitude of the electric field. As electric field lines indicate the direction of the resultant force acting on a charge, the lines cannot cross.

Suppose two point charges Q and Q' are separated by a distance r. The force on charge Q' is given by Coulomb's law:

$$F = \frac{QQ'}{4\pi\epsilon_0 r^2}$$

By using equation (13:3) the electric field can be found at the point in space where charge Q' is situated:

$$E = \frac{F}{Q'} = \frac{QQ'}{4\pi\epsilon_0 r^2 Q'} = \frac{Q}{4\pi\epsilon_0 r^2} \qquad (13:4)$$

Fig. 13:4 shows how the magnitude of the electric field varies with distance from the charge.

As stated previously, the electric field is a vector quantity. Therefore, in order to determine the electric field at a point in space in the vicinity of a number of point charges, it is necessary to use vector addition to determine the resultant field at that point.

Example 13:2

a) Calculate the magnitude of the electric field a distance of 30 cm from a point charge A of value 5×10^{-9} C. Draw a diagram showing the electric field lines in the vicinity of the charge.

b) Calculate the magnitude of the force on a charge of value 6×10^{-12} C placed at a distance of 30 cm from charge A.

Solution Refer to fig. 13:5, for the field lines.

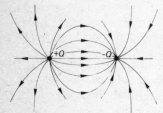

Fig. 13:3 Electric field line configuration for two point charges of opposite sign.

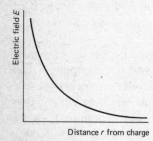

Fig. 13:4 Variation of electric field with distance from a positive point charge.

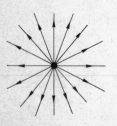

Fig. 13:5 Electric field lines radiating from a positive point charge.

a) The electric field at distance *r* from a point charge *Q* is given by equation (13:4):

$$E = \frac{Q}{4\pi\epsilon_0 r^2} = \frac{5\times10^{-9}}{4\pi\times8.85\times10^{-12}\times(30\times10^{-2})^2} = 500 \text{ N/C}$$

b) $F = EQ = 500\times6\times10^{-12} = 3\times10^{-9} \text{ N}$

13:5 Electric Potential

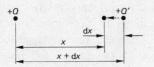

Fig. 13:6 Work done moving charge *Q'* closer to charge *Q*.

When two charges are a very great distance apart, the electrostatic force which each charge experiences is small. However if (for example) two positive charges are brought together, a force of repulsion must be overcome which increases as the distance between the charges decreases.

In moving a charge through a distance in the direction of the force, work is done (see section 4:2). Consider the situation described by fig. 13:6. The work done when the charge *Q'* is moved from a distance *x* + d*x* to a distance *x* from *Q* is −*F* d*x*, where *F* is the electrostatic repulsion between the two charges, d*x* is the very small distance through which *Q'* has moved, and we assume that *F* does not change over this distance. The minus sign appears because the charge *Q'* is moved in the opposite direction to the electrostatic force which acts on it.

The small amount of work done when the force *F* moves through a distance d*x* is d*W* given by

$$\mathrm{d}W = -F\,\mathrm{d}x$$

which using $F = EQ'$ becomes $\mathrm{d}W = -EQ'\,\mathrm{d}x$.

To find the total work done in bringing the charge *Q'* from an infinite distance from *Q* to some point, say a distance *r* from *Q*, we must integrate the above equation over that distance. In doing this we are summing the work done in moving the force *F* through successive intervals of d*x* from infinity up to a distance *r* from *Q*, taking into account that *F* is gradually changing as the charge is moved.

The electric field *E* a distance *x* from the charge *Q* can be found by using equation (13:4), i.e.

$$E = \frac{Q}{4\pi\epsilon_0 x^2}$$

Substituting this into the expression for d*W* gives

$$\mathrm{d}W = \frac{-QQ'\,\mathrm{d}x}{4\pi\epsilon_0 x^2} \tag{13:5}$$

Integrating gives the total work done in moving the charge *Q'* from infinity to *r*:

$$W = \frac{-QQ'}{4\pi\epsilon_0}\int_\infty^r \frac{\mathrm{d}x}{x^2}$$

gives $\quad W = \dfrac{QQ'}{4\pi\epsilon_0 r} \tag{13:6}$

Note The details of the integration can be omitted, it is the result that is important.

Equation (13:6) gives the work done *on* the charge Q' by some external agent.

The **electric potential** at a point is defined as the work done to bring a unit charge from infinity to that point in the electric field. Using equation (13:6)

$$V = \frac{W}{Q'} = \frac{Q}{4\pi\epsilon_0 r} \qquad (13:7)$$

where V is the electric potential. The unit of electric potential† is the VOLT, otherwise known as the joule per coulomb (abbreviated to J/C).

Both work and charge are scalar quantities and therefore potential is also a scalar quantity. It follows that, in calculations to determine the potential at a point in space due to a number of charges, we simply add together the contribution to the potential at that point due to the individual charges.

Equation (13:7) is applicable to a point charge and to a uniformly charged sphere of radius R, provided R is less than r, and r is measured from the centre of the sphere.

Example 13:3

Two positive point charges A and B carry charge 10^{-6} C and 2×10^{-6} C respectively.

a) If they are placed 10 cm apart, find the work done when another positive charge of value 10^{-7} C is brought up to a point midway between A and B.

b) Suppose that the 10^{-7} C charge is moved from its position in part *a*) till it is 1 cm from charge B. Calculate how much additional work is done.

Solution

a) Refer to fig. 13:7(a). The potential at P due to charge A, V_{PA}, can be found using equation (13:7):

$$V_{PA} = \frac{10^{-6}}{4\pi\epsilon_0(5\times10^{-2})} = 1.80\times10^5 \text{ V}$$

The potential at P due to charge B, V_{PB}, is given by

$$V_{PB} = \frac{2\times10^{-6}}{4\pi\epsilon_0(5\times10^{-2})} = 3.60\times10^5 \text{ V}$$

Therefore the potential at P due to both charges is

$$V_P = V_{PA} + V_{PB} = 5.4\times10^5 \text{ V}$$

(a)

(b)

Fig. 13:7

From equation (13:7) we have $V = W/Q'$ which gives $W = VQ'$ where, in this question, W represents the work done bringing the charge to point P. Substituting for V and Q' gives

$$W = 5.40\times10^5 \times 10^{-7} = 5.40\times10^{-2} \text{ J}$$

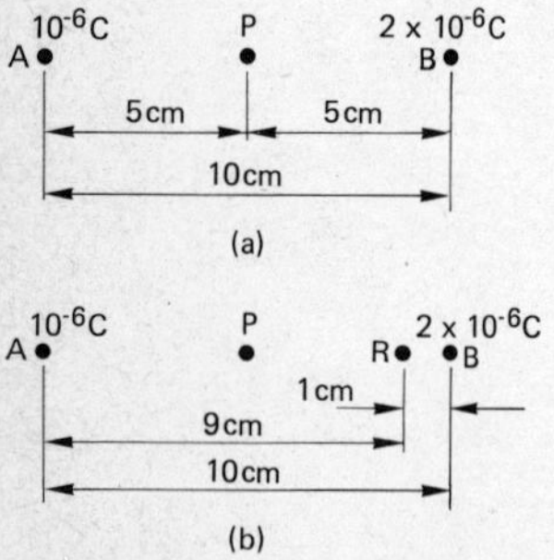

† Electric potential is often abbreviated to potential (the word "electric" is taken as read).

b) Refer to fig. 13:7(b). In order to determine the work done in moving the charge from P to R, we calculate the *difference* in potential between point P and R and multiply this by the value of the charge that is moved. That is,

Work done moving charge from P to R $= (V_R - V_P)Q$

The potential at R is

$$V_R = V_{RA} + V_{RB}$$

$$= \frac{10^{-6}}{4\pi\epsilon_0(9\times10^{-2})} + \frac{2\times10^{-6}}{4\pi\epsilon_0(1\times10^{-2})}$$

$$= (9.99\times10^4\,\text{V}) + (1.80\times10^6\,\text{V}) = 1.90\times10^6\,\text{V}$$

The difference in potential between points P and R (termed the **potential difference**) is

$$V_{PR} = V_R - V_P = 1.90\times10^6 - 5.40\times10^5 = 1.36\times10^6\,\text{V}$$

Therefore the work done moving the 10^{-7} C charge from P to R is

$$W = V_{PR}Q = 1.36\times10^6 \times 10^{-7} = 0.136\,\text{J}$$

13:6 Potential Difference and Lines of Equipotential

Part *b* of example 13:3 brings up an interesting question. Suppose that a charge is moved from one point in an electric field to another point at the *same* potential; how much work is done? The answer is that no work is done! Refer to fig. 13:8: if $V_X = V_Y$ then $V_X - V_Y = 0$. Consider placing a charge $+Q'$ at X. The direction of the electrostatic force acting on it is given by the direction of the electric field line at that point. If the charge is moved at right angles to the electric field direction, no work is done because there is no component of the electrostatic force acting on the charge in the direction of motion. A charge can be moved from X to Y without any work being done and in fact it can be shown that, no matter what path is taken from X to Y, the net work done on the charge is *zero*. Lines that intersect electric field lines at right angles have the property that each point on such a line is at the same potential and such lines are called **lines of equipotential**. Figs. 13:9 and 13:10 show the equipotential lines associated with a positive point charge, and two opposite charges in close proximity, respectively.

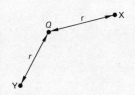

Fig. 13:8 Two arbitrary points a distance r from a point charge. Potential at X is V_X and the potential at Y is V_Y.

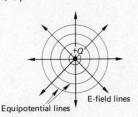

Equipotential lines   E-field lines

Fig. 13:9 Equipotential lines associated with a positive point charge.

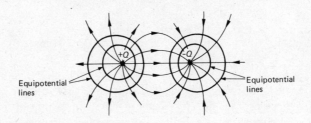

Equipotential lines   Equipotential lines

Fig. 13:10 Equipotential lines associated with two charges of opposite sign in close proximity.

Work is done when a charge moves through a potential difference. For convenience we often take the zero of potential to be at infinity in electrostatic calculations. However, when dealing with electric circuits, this process is less useful and we are often concerned with potential differences which are produced by, say, a battery. Nevertheless the principle is the same so that, if a charge Q moves between the two terminals of a battery, and the potential difference between the terminals is V, then the work done on the charge is $W = VQ$. We will return to the idea of potential difference in circuits in the next chapter.

13:7 Kinetic Energy of an Electron Moving through a Potential Difference

Suppose an electron moves through a potential difference V. The work done on the electron is given by $W = VQ$ where Q is the charge carried by the electron. Conventionally we represent the magnitude of the charge carried by the electron by the symbol e. The work done on the electron as it moves through the potential difference appears as energy of motion of the electron, i.e. kinetic energy. We have

$$Ve = \tfrac{1}{2}mv^2 \tag{13:8}$$

where m is the mass of the electron and v is its velocity.

Example 13:4

An electron in a vacuum tube is accelerated through a potential difference of 300 V. Calculate:
    a) The kinetic energy gained by the electron.
    b) The electron's final velocity.

Solution

a) Work done on the electron = kinetic energy gained by electron, and $W = Ve$. The magnitude of the charge carried by the electron is 1.6×10^{-19} C (see section 18.1)

$$W = 300 \times 1.6 \times 10^{-19} = 4.8 \times 10^{-17} \text{ J}$$

b) From equation (13:8),

$$\tfrac{1}{2}mv^2 = Ve \quad \text{so} \quad \tfrac{1}{2}mv^2 = 4.8 \times 10^{-17} \text{ J}$$

The mass of the electron is 9.1×10^{-31} kg. Therefore rearranging the above equation gives

$$v^2 = \frac{2 \times 4.8 \times 10^{-17}}{9.1 \times 10^{-31}} \quad \text{so} \quad v = 1.03 \times 10^7 \text{ m/s}$$

13:8 Relationship between Potential and the Electric Field

A superficial glance at equations (13:4) and (13:7), which are the expressions for the electric field and the potential a distance r from a point charge, would indicate that they are very similar. In fact if we differentiate equation (13:7) with respect to r we get

$$\frac{dV}{dr} = \frac{-Q}{4\pi\epsilon_0 r^2}$$

Comparing this with the expression for E, it can be seen that

$$E = -\frac{dV}{dr} \qquad (13:9)$$

A full explanation of this equation is beyond the scope of this book. However it shows that an alternative unit for the electric field is the volt per metre (abbreviated to V/m).

Equation (13:9) can be used to determine the electric field at a point in space in many practical situations. One of the most important cases is that of the electric field between two parallel metal plates that carry a charge. In this situation the field lines are all parallel and the field is described as *uniform*. If the potential between the plates is V and the distance between the plates is d, the magnitude of the electric field between the plates in this case is

$$E = \frac{V}{d} \qquad (13:10)$$

13:9 Capacitance

It can be shown experimentally that, as the charge on a conductor increases, the potential at the surface of the conductor also increases. Writing this mathematically

$$V \propto Q$$

This leads to the definition of a quantity called the capacitance of a conductor. If a charge Q raises the potential of a conductor by an amount V, the **capacitance** of the conductor is given by

$$C = \frac{Q}{V} \qquad (13:11)$$

The unit of capacitance is the FARAD (symbol F). The farad is defined as follows:

If a conductor changes its potential by one volt when one coulomb of charge is placed upon it, then the capacitance of that conductor is one farad.

In practical situations a capacitance of 1 F is rare. Much more common units are microfarads μF (10^{-6} F), nanofarads nF (10^{-9} F), and picofarads pF (10^{-12} F).

Take as an example the capacitance of a spherical conductor. The potential at the surface of a spherical conductor of radius a, carrying a charge Q, is given by $V = Q/4\pi\epsilon_0 a$. Substituting this into equation (13:11) gives

$$C = \frac{Q}{V} = \frac{Q}{\left(\dfrac{Q}{4\pi\epsilon_0 a}\right)} = 4\pi\epsilon_0 a \qquad (13:12)$$

Capacitors are found in a great many electrical circuits, the most common form of capacitor in these situations being the parallel plate capacitor.

13:10 Parallel Plate Capacitor

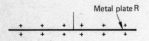

Fig. 13:11 Metal plate carrying a positive charge.

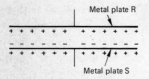

Fig. 13:12 Uncharged metal plate S brought up to a charged metal plate R.

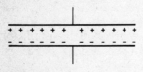

Fig. 13:13 Usual distribution of charge on a parallel plate capacitor.

A **parallel plate capacitor** consists of two parallel metal plates which are separated by an insulator. Initially consider the plates separated by an air gap. Fig. 13:11 shows one plate carrying a positive charge. The potential of plate R will be quite high due to the positive charge that it possesses. Now bring plate S which is uncharged, close to plate R. The effect of the charge on R will be to separate the charge on S as shown in fig. 13:12. Electrons are attracted to the upper surface of S, leaving the lower surface deficient in electrons. The potential at the surface of R will now be lower due to the presence of the negative charge close by on S. The potential difference between the two plates has decreased but the charge stored has not changed; therefore the capacitance has increased. Further increase in the capacitance can be produced by eliminating the positive charge on S which can be done by connecting S to earth momentarily. Fig. 13:13 shows the charge on the plates after this has been done. Electrons have been attracted from the earth and have cancelled out the positive charge on S.

The above situation described is quite artificial. Normally a parallel plate capacitor is given a charge by connecting it to a source of potential difference, such as a battery, which has the effect of moving the electrons from one plate onto the other.

It can be shown that the formula for the capacitance of a parallel plate capacitor is given by

$$C = \frac{\epsilon_0 A}{d} \qquad (13:13)$$

where A is the common area of overlap between the two plates and d is the distance between the plates.

If a non-conducting material, known as a **dielectric**, is placed between the plates of a parallel plate capacitor, a further increase in capacitance can be produced. This arises from the fact that, although the electrons are not free to move in the dielectric, the electron clouds can become

Fig. 13:14 Distortion of electron clouds in a dielectric placed between plates of a capacitor.

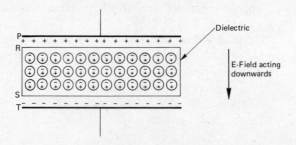

distorted (see fig. 13:14). The electron clouds surrounding each atom shift slightly, relative to their nuclei. This is due to the electric field existing between the plates which acts to attract the negative charge (on the electrons) towards the positive plate, and the positive charge (on the nuclei) towards the negative plate. At the upper surface R, the negative charge tends to lower the potential of the plate P. Similarly the positive charge at S will tend to raise the potential at T. Thus the potential difference (p.d.) is decreased but the charge on the plates is unchanged.

The capacitance of a parallel plate capacitor containing a dielectric is given by

$$C_D = \frac{\epsilon A}{d} \qquad (13{:}14)$$

where ϵ is called the permittivity of the dielectric.

The **relative permittivity** of a dielectric, ϵ_r, is defined as the ratio of the capacitance of a capacitor containing the dielectric to the capacitance of the capacitor without the dielectric.

$$\epsilon_r = \frac{C_D}{C} = \frac{\left(\dfrac{\epsilon A}{d}\right)}{\left(\dfrac{\epsilon_0 A}{d}\right)}$$

which gives $\quad \epsilon_r = \dfrac{\epsilon}{\epsilon_0} \qquad (13{:}15)$

Note that ϵ and ϵ_0 have units F/m but ϵ_r has no units since it is simply a ratio.

13:11 Combination of Capacitors

As stated previously capacitors are to be found in a great many electrical (and electronic) circuits. Fig. 13:15 shows a selection of commonly available capacitors. Capacitors can be combined together to produce a variety of effective capacitances in a circuit if the desired value of capacitance is not available.

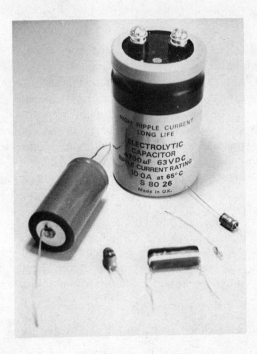

Fig. 13:15 Selection of commercially available capacitors.

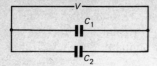

Fig. 13:16 Capacitors in parallel combination.

1 Fig. 13:16 shows two capacitors in a **parallel** combination. If a potential difference V, say supplied by a battery, is placed between the ends of the combination, it follows that the p.d. across each capacitor will be the same, i.e.

$$V = V_1 = V_2$$

where V_1 and V_2 are the p.d.s across capacitors C_1 and C_2 respectively.

The potential across C_1 is given by

$$V_1 = \frac{Q_1}{C_1} \qquad \text{similarly} \qquad V_2 = \frac{Q_2}{C_2}$$

therefore $Q_1 = C_1 V_1$ and $Q_2 = C_2 V_2$

where Q_1 and Q_2 are the charges stored on capacitors C_1 and C_2 respectively.

$$\text{Now} \quad V = \frac{Q}{C_P} \quad \text{so that} \quad Q = C_P V$$

where Q is the total charge stored on both capacitors and C_P is the effective capacitance of the capacitors in parallel.

We have $Q = Q_1 + Q_2$

Substituting for Q, Q_1 and Q_2 from above leads to

$$C_P V = C_1 V_1 + C_2 V_2$$

and as $V = V_1 = V_2$ we can cancel the Vs to give

$$C_P = C_1 + C_2 \tag{13:16}$$

This formula can be extended to any number of capacitors in parallel.

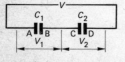

Fig. 13:17 Capacitors in series combination.

2 Fig. 13:17 shows two capacitors C_1 and C_2 connected in a **series** combination. If a charge $+Q$ accumulates on side A of capacitor C_1 when the battery is connected, then charge $-Q$ must be held on plate B. Similarly the charge on plate D of capacitor C_2 is the same as that on plate A of capacitor C_1 but of the opposite sign (electrons have moved from plate A to plate D). Therefore the charge on plate C must be $+Q$. We conclude that when capacitors are connected in series the *charge* on each capacitor is the same.

Using Q to represent the charge on each capacitor, we have

$$V_1 = \frac{Q}{C_1} \qquad V_2 = \frac{Q}{C_2} \qquad V = \frac{Q}{C_S}$$

where C_S is the effective capacitance of the capacitors in this combination. The sum of the potential differences across each capacitor gives the total potential difference across the combination

$$V = V_1 + V_2$$

Now we replace V, V_1 and V_2 from the above relations to give

$$\frac{Q}{C_S} = \frac{Q}{C_1} + \frac{Q}{C_2} \qquad \text{and cancelling } Q \text{ gives}$$

$$\frac{1}{C_S} = \frac{1}{C_1} + \frac{1}{C_2} \qquad\qquad (13{:}17)$$

This formula can be extended to any number of capacitors in series.

It can be seen that the greater the number of capacitors placed in parallel, the *greater* is the effective capacitance. However, when capacitors are placed in series, the total effective capacitance is *less* than any one of the capacitors making up the series combination.

Example 13:5

Between the terminals of a dry battery there is a p.d. of 1.5 volts. The battery is connected to a 5 μF capacitor.

a) Calculate the charge stored on the plates of the capacitor. The battery is disconnected and the charged 5 μF capacitor is connected in parallel to an uncharged 10 μF capacitor. Calculate

b) The potential across the capacitors.

c) The charge on each capacitor.

Solution

a) Rearranging equation (13:11) gives

$$Q = VC = 1.5 \times 5 \times 10^{-6} = 7.5 \ \mu\text{C}$$

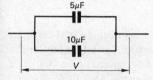

Fig. 13:18 Combination of capacitors

b) Refer to fig. 13:18. The total charge stored will not change when the two capacitors are connected together although some charge will flow from the 5 μF capacitor to the 10 μF capacitor.

Total charge $Q = Q_1 + Q_2 = 7.5 \ \mu\text{C}$

Now $C_P = C_1 + C_2 = 10 \ \mu\text{F} + 5 \ \mu\text{F} = 15 \ \mu\text{F}$

It follows that

$$V = \frac{Q}{C_P} = \frac{7.5 \times 10^{-6}}{15 \times 10^{-6}} = 0.5 \text{ V}$$

c) $Q_1 = C_1 V = 0.5 \times 5 \times 10^{-6} = 2.5 \ \mu\text{C}$

$Q_2 = C_2 V = 0.5 \times 10 \times 10^{-6} = 5.0 \ \mu\text{C}$

13:12 Energy Stored in Charged Capacitor

Consider a capacitor of capacitance C carrying an initial charge q and a corresponding p.d. across the capacitor of V. Suppose a further small amount of charge dq is to be placed on the plates. Then work must be done to achieve this. The small amount of work done, which appears as energy stored in the capacitor, is given by

$$dW = V \, dq$$

If the capacitor is initially uncharged, it is possible to calculate the energy stored when a total charge Q is placed on the plates of the capacitor by integrating the above equation.

This gives

$$W = \int_0^Q V \, dq \qquad \text{but } V = \frac{q}{C} \text{ and therefore}$$

$$W = \frac{1}{C} \int_0^Q q \, dq = \frac{1}{2} \frac{Q^2}{C}$$

If Q is replaced by CV we can write

$$W = \tfrac{1}{2} CV^2 \qquad\qquad\qquad\qquad\qquad\qquad (13{:}18)$$

Example 13:6

A capacitor of capacitance 5 pF is charged to a p.d. of 100 V. Calculate the energy stored in the capacitor.

Solution Using equation (13:18),

$$W = \tfrac{1}{2} CV^2 = \tfrac{1}{2} \times 5 \times 10^{-12} \times (100)^2 = 2.5 \times 10^{-8} \, J$$

Exercises

13.1 If a toothbrush is rubbed vigorously with a piece of fur it is able to attract small pieces of paper. Explain why this happens.

13.2 Describe an experiment which you could perform to determine the sign of the charge on a body that has been charged by friction.

13.3 Two point charges of values 3×10^{-7} C and 4×10^{-8} C are a distance of 5 cm apart in a vacuum. Show on a diagram the force acting on each charge and calculate the magnitude of the force. Suppose the 3×10^{-7} C charge is replaced by a charge of -3×10^{-7} C. Explain how this affects the force acting on the 4×10^{-8} C charge.

13.4 Give the value of the electric field strength a distance of 10 cm from a negative point charge of magnitude 4×10^{-5} C. A second charge of value 8×10^{-12} C is placed a distance of 10 cm from this charge. Calculate the magnitude of the force that it experiences.

13.5 Two charges A and B have value 10^{-8} C and 3×10^{-6} C respectively. Initially the two charges are a distance of 5 cm apart. Calculate the work done when charge A is moved a distance of 50 cm from charge B.

13.6 Define potential. Calculate the work done on a unit charge to bring it from infinity to a distance of 30 cm from a point charge of 4×10^{-11} C.

13.7 An insulated body has been struck by lightning and its potential has been raised to 1.5 MV. Calculate how much work would have to be done to place a further charge of 10^{-2} C onto the body (assume this charge is brought from an infinite distance away from the body).

13.8 Define capacitance. A hollow metal sphere of radius 2 m is charged to a potential of 3 V. Calculate the charge it carries.

13.9 If the earth can be considered as a spherical conductor of radius 6400 km, estimate its capacitance.

13.10   A parallel plate capacitor has the following specification. Area of overlap of plates is $20\,cm^2$. Distance between the plates is $0.1\,mm$. The dielectric sandwiched between the plates has a relative permittivity of 11. Calculate the capacitance of this arrangement.

13.11   Two capacitors of value $3\,\mu F$ and $7\,\mu F$ are connected in series. If the p.d. across the combination is $3\,V$, calculate
    a) The effective capacitance of this arrangement.
    b) The charge on each capacitor.
    c) The potential difference across each capacitor.

13.12   A capacitor of value $3\,mF$ is charged to a potential difference of $50\,V$. This capacitor is then connected in parallel to a $500\,\mu F$ capacitor (which was originally uncharged). Calculate
    a) The p.d. across the combination.
    b) The final charge on each capacitor.

13.13   Which of the following are scalar and which are vector quantities?
    a) Charge   *d*) Work
    b) Force   *e*) Capacitance
    c) Potential   *f*) Electric Field.

13.14   A p.d. of $12\,V$ is placed across two capacitors of value $3\,pF$ and $10\,pF$ in series combination. Calculate the energy stored in each capacitor.

14 Moving Charges and Circuits

The concepts of electric field and potential introduced in the last chapter were derived from an investigation of the properties of stationary charges. Our discussion is no longer limited to stationary charges in free space and on the surface of conductors, but is extended to investigate the consequences of setting up an electric field *within* a conductor.

14:1 Electric Current

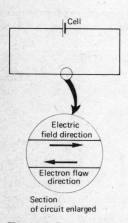

Fig. 14:1 Force on electrons due to an electric field within a conductor.

As we have already seen, there are electrons within conductors that are relatively weakly bound to their parent atoms. In good conductors such as copper there are of the order of 10^{29} free electrons per cubic metre. Suppose a potential difference is placed across a conductor, for instance a length of copper wire.† The potential difference sets up an electric field in the copper; free electrons within the copper experience a force and are compelled to move by that force. Fig. 14:1 represents a simple circuit consisting of a source which is able to drive electrons through a conductor. The positively charged nucleus of each copper atom and the electrons close to the nucleus also experience a force due to the electric field, but they are firmly bound and so are not able to move freely under the influence of the electric field.

The movement of charge through a material constitutes an **electric current**. The unit of current is the ampere (usually abbreviated to the amp.). The **ampere** is related to the coulomb as follows:

If one coulomb of charge passes a point in a circuit each second, then the current passing that point is one ampere. Beware that this is *not* a definition of the ampere; for that we must relate charge flow to more fundamental quantities such as force and length. This will be dealt with in section 15:9.

We may write

$$I = \frac{Q}{t} \tag{14:1}$$

where I is the current (in amps), Q is the amount of charge passing a point in the circuit in a time t.

---

† The potential difference could be supplied by a battery but it is strongly recommended that the student does not connect a length of copper wire across the terminals of a battery, as this will cause the battery to discharge very quickly.

14:2 Conventional Current and the Direction of Electron Flow

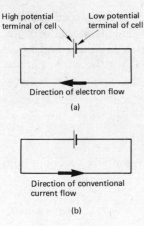

High potential terminal of cell

Low potential terminal of cell

Direction of electron flow

(a)

Direction of conventional current flow

(b)

The electrons carry charge around a circuit. The electrons move from a position of low potential to a position of high potential as shown in fig. 14:2(a). Before the existence of the electron was confirmed, it was assumed that the charge carriers which moved in an electric circuit carried a positive charge. Such a charge carrier would move from a position of high potential to a position of low potential shown in fig. 14:2(b). This is taken to be the direction of *conventional current flow*. When the direction of current is given, it is implied, unless stated otherwise, that conventional current flow is being considered and the direction of the flow of electrons is ignored.

Fig. 14:2 Direction of electron flow and conventional current flow compared.

14:3 Ohm's Law and Resistance

If the potential difference across the ends of a conductor is increased, the current flowing through the conductor also increases. It was found experimentally that for many conductors the current increases in direct proportion to the increase in p.d. Another way of saying this is that

$$\frac{V}{I} = \text{a constant}$$

where V is the p.d. across the conductor, often loosely referred to as the "voltage", and I is the current flowing through the conductor. This is known as **Ohm's law**. Ohm's law holds accurately for many conductors so long as the physical conditions do not change. For example, the conductor should not be stretched, compressed or subject to an increase in temperature.

The ratio of the p.d. across a conductor to the current flowing through the conductor is called the **resistance** of the conductor (symbol R). This can be written

$$\frac{V}{I} = R \tag{14:2}$$

The unit of resistance is the OHM, symbol Ω.

If a current of one ampere flows through a conductor while a p.d. of one volt is maintained between its ends, the conductor has a resistance of one ohm.

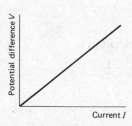

Potential difference V

Current I

Fig. 14:3 Graph of p.d. versus current for a conductor that obeys Ohm's law.

Fig. 14:3 shows a graph of p.d. against current for a conductor. The straight line indicates that the ratio V/I is a constant and therefore this conductor obeys Ohm's law. Anything other than a straight line would mean that Ohm's law is *not* obeyed.

14:4 Resistivity

The resistance of a conductor can be related to its dimensions. For example, if the path along which the charges have to travel is doubled, then the resistance is also doubled, i.e.

$$R \propto L$$

where L is the length of the conductor. Increasing the cross-sectional area of a conductor decreases its resistance. A useful analogy is that of water flowing through pipes. The greater the diameter of the pipe, and hence the greater the cross-sectional area, the easier it is for the water to flow. Similarly the greater the cross-sectional area of the wire, the easier it is for the electrons to flow. We may write

$$R \propto \frac{1}{A}$$

Where A is the cross-sectional area of the wire. Taking into account both the length of the wire and the cross-sectional area, we have

$$R \propto \frac{L}{A}$$

The proportionality sign can be replaced by a constant to give

$$R = \rho \frac{L}{A} \tag{14:3}$$

ρ is the **resistivity** and is constant for a particular material (see table 14:1). If R is measured in ohms, L in metres and A in square metres, then the units of resistivity can be found by rearranging equation (14:3):

$$\rho = \frac{RA}{L} = \frac{\text{ohm (metre)}^2}{\text{metre}} = \text{ohm-metres } (\Omega\,\text{m})$$

Table 14:1 Resistivity of some common materials at room temperature.

MATERIAL	ρ (Ωm)	USE
Copper	1.7×10^{-8}	Connecting wires
Gold	2.4×10^{-8}	Low-resistance contacts
Manganin	4.0×10^{-7}	Resistance wire
Nichrome	1.0×10^{-6}	Resistance wire
Germanium	4.3×10^{-1}	Transistors
Silicon	2.6×10^{3}	Transistors
Glass	$> 10^{10}$	Insulation
Mica	$> 10^{11}$	Dielectric (in capacitors)
Rubber	$> 10^{12}$	Insulation

14:5 The Heating Effect of a Current

A conductor can be considered as consisting of positive ions (positive nucleus plus tightly bound electrons) in fixed positions within a lattice, surrounded by a "sea" of free electrons. The electrons are moving at random and there is no net motion of the electrons in any direction until a p.d. is applied to the material. When this happens the electrons accelerate due to the electric field created. However the electrons do not continue to accelerate indefinitely. Electrons collide with the lattice ions and in doing so lose the energy they have gained from the electric field. The energy that is passed onto the conductor raises the temperature of the conductor. When a charge of Q coulombs moves through a p.d. of V volts, the amount of energy "consumed" is given by $W = VQ$ (see section 13:5).

If Q coulombs of charge pass a point in a circuit in t seconds, the rate at which energy is converted is given by

$$\frac{W}{t} = \frac{VQ}{t}$$

Q/t can be replaced by I from equation (14:1), which gives

$$\frac{W}{t} = VI \qquad (14:4)$$

The rate at which energy is converted is known as the **power** (symbol P). This is the energy converted each second from, say, chemical energy in a battery, to thermal energy in the conductor. If we write W/t as P then

$$P = VI \qquad (14:5)$$

Substitution for V from equation (14:2) gives

$$P = I^2 R \qquad (14:6)$$

Example 14:1

1 A resistor of resistance $10\,\Omega$ is connected to a battery which provides a p.d. between its terminals of $2\,V$. Calculate
    a) The current flowing through the resistor.
    b) The energy dissipated per second in the resistor.
    c) The total energy delivered to the resistor in one minute.

Solution

a) Using equation (14:2) $I = \dfrac{V}{R} = \dfrac{2}{10} = 0.2\,A$

b) The energy dissipated per second in the resistor is given by equation (14:5):

    $P = VI = 2 \times 0.2 = 0.4\,W$

c) Power is the rate at which work is done, i.e. $P = \dfrac{W}{t}$.

    Rearranging gives $W = Pt = 0.4 \times 60 = 24\,J$

2 A current of .5 A flows in a wire of resistivity $2.5 \times 10^{-6}\,\Omega\,m$. If a p.d. of $10\,V$ is maintained between the ends of the wire and the length of the wire is $5\,m$ calculate the diameter of the wire.

Solution The resistance of the wire is

$$R = \frac{V}{I} = \frac{10}{5} = 2\,\Omega$$

Rearranging equation (14:3) gives

$$A = \frac{\rho L}{R} = \frac{2.5 \times 10^{-6} \times 5}{2} = 6.25 \times 10^{-6}\,\text{m}^2$$

Area of cross-section $A = \pi r^2$, where r is the radius of the wire, so that

$$r = \sqrt{\frac{A}{\pi}} = \sqrt{\frac{6.25 \times 10^{-6}}{\pi}} = 1.41 \times 10^{-3}\,\text{m}$$

The wire diameter $d = 2r = 2.82 \times 10^{-3}\,\text{m}$

14:6 The Effect of Temperature upon Resistance

Atoms (or ions) within a solid are constantly vibrating about a mean position. As the temperature of the solid increases, so the amplitude of the vibration also increases. In our model of electrons drifting through a conductor under the influence of an electric field, the increased vibrations increase the likelihood of a collision occurring with an electron. This results in an increase in the resistance of the conductor. For many conductors the resistance over a limited temperature range can be described by the following relation:

$$R_\theta = R_0 + \text{a factor depending on } \theta$$

where R_θ is the resistance at the temperature θ, and R_0 is the resistance at 0°C. The full relationship is as follows:

$$R_\theta = R_0 + R_0\alpha\theta$$

or $\quad R_\theta = R_0(1 + \alpha\theta)$ \hfill (14:7)

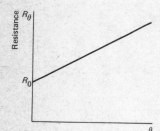

α is known as the **temperature coefficient of resistance** (units: per °C). Fig. 14:4 shows a typical graph of R against θ for a conductor.

Fig. 14:4 Resistance versus temperature graph for a conductor obeying relation (14:7).

Equation (14:7) indicates that, for any conductor, Ohm's law is not strictly obeyed if the temperature increases, unless $\alpha = 0$. However in many cases θ will be a matter of a few degrees Celsius and as α is small, typically of the order $10^{-3}/°C$ to $10^{-4}/°C$, the additional factor is very small and in the majority of cases can be ignored.

14:7 Electromotive Force and Potential Difference

Any source which can drive current around a circuit has associated with it an electromotive force (abbreviated as e.m.f.). A common source of e.m.f. is the dynamo of a generator in which mechanical energy is converted into electrical energy. The dynamo will be dealt with later, here we will restrict our discussion to batteries and cells.

The **e.m.f. of a cell** is the amount of energy converted from chemical energy (stored in the cell) to electrical energy (associated with the charge flowing through the circuit) per unit charge flowing. Referring to fig. 14:5, if 10 J of energy are converted from chemical to electrical energy when a

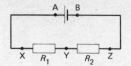

Fig. 14:5 Explanation of electromotive force and potential difference.

Fig. 14:6 Graphical representation of the change in potential around an electric circuit.

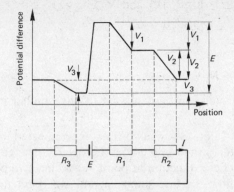

charge of 2 C flows from A through resistors R_1 and R_2 to B, the energy converted per coulomb is therefore $10/2 = 5$ J/C, which is the e.m.f. of the cell. Since the J/C is the volt, the e.m.f. is measured in volts.

The difference in potential between A and B is called the e.m.f. of the cell. It should be emphasised that an e.m.f. is associated with a *source of energy* in a circuit. When we speak of a **potential difference** we can consider **any two points in a circuit**, for example X and Y or Y and Z in fig. 14:5. The potential difference between these points is the quantity of electrical energy converted to other forms of energy (e.g. heat and light) when a charge of one coulomb flows between the two points.

In short, *the e.m.f. tells us about the energy converted in the source to electrical energy* and *the potential difference tells us about the energy converted from electrical energy to other forms of energy.*

A graph showing how potential varies with position round a circuit is shown in fig. 14:6. The potential decreases from its highest value at the positive side of the cell to its lowest value at the negative side of the cell. If the current flowing through the circuit is known, then the p.d. across each resistor can be determined using $V = IR$. The sum of the p.d.s across each resistor can be seen to be equal to the e.m.f. of the cell. We may write

$$E = V_1 + V_2 + V_3 \qquad (14:8)$$

where E is the e.m.f. of the cell.

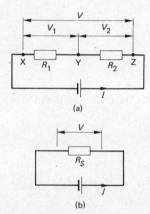

Fig. 14:7 (a) Two resistors in series combination; (b) equivalent circuit.

14:8 Combination of Resistors

1 Fig. 14:7(a) shows two resistors connected in **series**. The resistances of R_1 and R_2 combine and can be represented by another resistor R_S of equivalent resistance, as shown in fig. 14:7(b).

The same current I flows through R_1 and R_2, and using equation (14:2),

$$V_1 = IR_1 \quad \text{and} \quad V_2 = IR_2$$

The p.d. across the equivalent resistor R_S is V, where $V = IR_S$.

The p.d. between X and Z is the sum of the p.d.s between X and Y and Y and Z, i.e.

$$V = V_1 + V_2$$

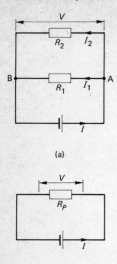

(a)

(b)

Fig. 14:8 (a) Two resistors in parallel combination; (b) equivalent circuit.

Substituting for V, V_1 and V_2 gives

$$IR_S = IR_1 + IR_2$$

Therefore,

$$R_S = R_1 + R_2 \qquad (14:9)$$

This formula can be extended to any number of resistors in series.

2 Fig. 14:8(a) shows two resistors connected in parallel. The potential V across R_1 is the same as that across R_2. The current passing through the cell reaches a junction at point A and has two possible paths along which it can flow. In fact some current will pass through each resistor. For convenience make the current through R_1 equal to I_1 and the current through R_2 equal to I_2. At junction B, I_1 and I_2 join together again and go on to pass through the cell.

The sum of the currents through R_1 and R_2 must be equal to the current flowing through the cell. Therefore

$$I = I_1 + I_2$$

Using equation (14:2),

$$I_1 = \frac{V}{R_1} \quad \text{and} \quad I_2 = \frac{V}{R_2}$$

The resistors in parallel can be represented by one equivalent resistor R_P, as shown in fig. 14:8(b). From equation (14:2) again,

$$I = \frac{V}{R_P}$$

As stated, $I = I_1 + I_2$, and therefore substituting for I, I_1 and I_2 gives

$$\frac{V}{R_P} = \frac{V}{R_1} + \frac{V}{R_2}$$

i.e. $\quad \dfrac{1}{R_P} = \dfrac{1}{R_1} + \dfrac{1}{R_2} \qquad (14:10)$

This formula can be extended to any number of resistors in parallel.

Example 14:2

1 Consider the circuit shown in fig. 14:9. Calculate:
  a) The current flowing through the circuit.
  b) The p.d. across the $2\,\Omega$ resistor.
  c) The energy dissipated per second in the $3\,\Omega$ resistor.

Solution
a) The total resistance in the circuit is found by summing the two resistances as given by equation (14:9):

$$R_S = R_1 + R_2 = 2 + 3 = 5\,\Omega$$

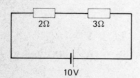

2Ω $\qquad$ 3Ω

10V

Fig. 14:9

To find the current, using equation (14:2):

$$I = \frac{V}{R_S} = \frac{10}{5} = 2\,\text{A}$$

b) $V_1 = IR_1 = 2 \times 2 = 4\,\text{V}$

c) The energy dissipated per second can be found using equation (14:6):

$$P = I^2R = 2^2 \times 3 = 12\,\text{W}$$

2 Consider the circuit shown in fig. 14:10. Calculate:
a) The p.d. across the resistors.
b) The current through the 2 Ω resistor.
c) The current through the 3 Ω resistor.

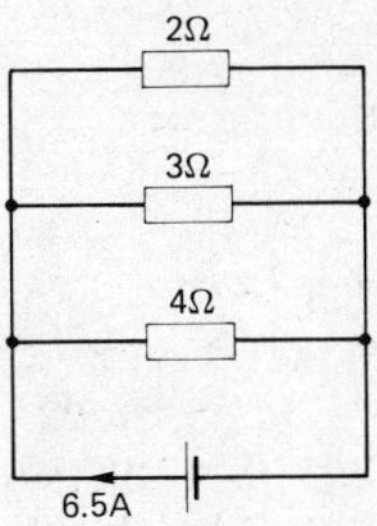

Solution
a) Calculate the effective resistance of the three resistors in parallel using equation (14:10):

$$\frac{1}{R_P} = \frac{1}{2} + \frac{1}{3} + \frac{1}{4} = \frac{6+4+3}{12} = \frac{13}{12}$$

Therefore $R_P = \frac{12}{13}\,\Omega$

The p.d. across the combination is given by

$$V = IR_P = \frac{6.5 \times 12}{13} = 6\,\text{V}$$

b) The current through the 2 Ω resistor is given by

$$I = \frac{V}{2} = \frac{6}{2} = 3\,\text{A}$$

Similarly current through the 3 Ω resistor is given by

$$I = \frac{V}{3} = \frac{6}{3} = 2\,\text{A}$$

Fig. 14:10

14:9 The Potential Divider

A number of resistors in series act as a **potential divider**. If, for example, a p.d. of 100 V is placed across the three resistors shown in fig. 14:11, the potential difference across each resistor can be found by first calculating the current through the resistors, then multiplying it by the resistance of each resistor. The ratio of the potential difference across each resistor is in the same ratio as the resistances. A potential divider can be useful when the required potential difference is not available. By placing two or more resistors in series, the required potential difference can be "picked off" across one of the resistors in the chain.

Fig. 14:11 The potential divider.

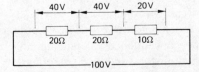

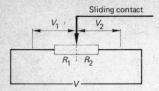

Fig. 14:12 The rheostat (schematic).

The **rheostat** shown in fig. 14:12 can act as a more flexible potential divider. It consists of a long coil of bare wire. A p.d. is placed between the ends of the wire and a contact can be made to any part of the wire via a movable metal brush. By adjusting the position of the sliding contact, the p.d.s V_1 and V_2 can be varied in the ratio $R_1 : R_2$.

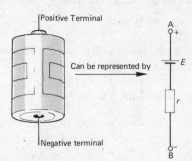

Fig. 14:13 Representation of internal resistance of a cell.

14:10 Internal Resistance

A voltmeter (see section 15:14) can be used to measure p.d.s. Can it therefore be used to measure e.m.f.s directly? As far as cells and batteries are concerned, the answer is a qualified "no" and this is due to the **internal resistance** associated with these sources of e.m.f. Due to its physical composition, every battery or cell has an internal resistance, which we can think of as being within the battery between the two exposed terminals as shown in fig. 14:13.

A voltmeter (which has a high resistance) when connected between terminals A and B of the battery will allow a small current to flow. If a current I flows, then there must be a potential difference across the internal resistance of value Ir, where r is the internal resistance of the battery. It follows that the voltmeter would not read E, the e.m.f. of the battery, but $(E - Ir)$. [Ir is sometimes called the "lost volts".] If the current I could be reduced to zero, then the voltmeter would read the e.m.f. of the battery exactly. This leads to an alternative definition of e.m.f. as the p.d. between the terminals of a battery when no current is flowing (often referred to as *open-circuit conditions*).

14:11 The Poten-tiometer

One instrument used to measure p.d. is the **potentiometer**. The instrument at its most simple consists of a long piece of resistance wire of uniform diameter across which there is a known p.d. (Refer to fig. 14:14.) The driving cell is typically a 2-volt accumulator which can maintain a large current for an extended period. The wire extending from A to B is, typically, manganin whose resistance does not change appreciably if the temperature of the wire increases.

In order to understand the principle of operation of the potentiometer, we consider the consequences of placing two cells in a circuit such that the p.d.s across their terminals are orientated as shown in fig. 14:15. Cell

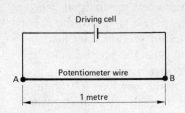

Fig. 14:14 The potentiometer.

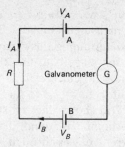

Fig. 14:15 Principle of operation of the potentiometer.

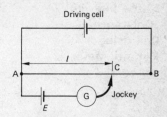

Fig. 14:16 Potentiometer arrangement used to measure an unknown p.d.

A is trying to drive current I_A around the circuit in an anticlockwise direction, while cell B is trying to drive current I_B around the circuit in a clockwise direction. If V_A is greater than V_B, then the current will flow anticlockwise, but clockwise if V_B is greater than V_A.

If $V_A = V_B$, *no current* will flow and the galvanometer† will register zero. This is important as it is the principle upon which the potentiometer operates.

Fig. 14:16 shows an arrangement that can be used to measure an unknown p.d. Suppose the p.d. between the ends of the metre wire is 2 V. The potential decrease per centimetre is therefore $2/100 = 0.02$ V/cm. One terminal of a cell of unknown e.m.f. is connected to A. (Note that the positive terminal of the unknown cell is connected to the same point as the positive terminal of the driving cell.) The other side of the cell of unknown e.m.f. is connected to a terminal of the sensitive galvanometer. The other terminal of the galvanometer is connected to a "jockey" which is a piece of metal with a blade that is used to make contact at any point along the potentiometer wire.

The jockey is moved carefully along the wire until a point is reached where the p.d. between A and C is equal to the unknown e.m.f. and the galvanometer registers zero, i.e. no current flows. This is termed the "balance-point". If the distance l between A and C is measured, the e.m.f. of the cell can be calculated.

$$E = \text{potential decrease per cm‡} \times \text{balance length}$$

E is the unknown e.m.f. and the balance length is l.

In our example where the potential decrease per cm is 0.02 we have $E = 0.02 \times l$. If $l = 75$ cm then the unknown e.m.f. would be

$$E = 0.02 \times 75 = 1.5 \text{ V}$$

---

† The galvanometer is an instrument which is used to detect and measure very small currents. The details of its construction will be dealt with in section 15:12.
‡ The potential decrease per cm is often referred to as the "volts-drop per cm."

14:12 Measurement of the Internal Resistance of a Cell using a Potentiometer

As stated in section 14:10 every cell has an internal resistance which cannot be measured directly. The potentiometer provides an elegant method of determining this quantity. There are two stages to the experiment. The first stage is to find the balance point produced by the cell in question when no current flows through the cell. The second stage is to allow current to flow and to find the new balance point.

Fig. 14:17(a) shows the first stage. Initially the key K is left open and the balance point l is found. The potential decrease per cm is given by E/l. No current passes through r and so there is no p.d. across it.

Fig. 14:17 Potentiometer used to measure the internal resistance of a cell.

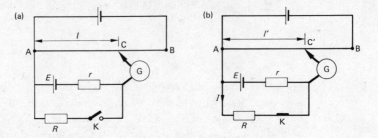

Fig. 14:17(b) shows the second stage. The circuit containing R is completed by closing the key K. Now current flows through r and R. We have

$$E = I(r + R) \qquad (14:11)$$

A new balance point C' is found and the distance l' measured. It can be seen that the p.d. across the standard resistor R is balanced against the p.d. across the length of wire l'.

If the p.d. across R is V, then $V = IR$. The balance length is l', therefore,

$$IR = \frac{E}{l} l' \qquad (14:12)$$

From equation (14:11),

$$I = \frac{E}{r + R}$$

which on substitution into equation (14:12) gives

$$\frac{El'}{l} = \frac{ER}{r + R}$$

Dividing both sides by E and rearranging gives

$$r = R\left(\frac{l}{l'} - 1\right) \qquad (14:13)$$

As l and l' can be measured and R is a standard resistance, r can be determined.

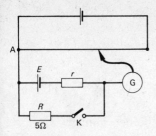

Fig. 14:18

Example 14:3

A circuit is set up as shown in fig. 14:18. When the key is left open, the balance point is found at 78.1 cm from A. With the key closed, the balance point is found at 50.5 cm from A. Explain why the differing balance points occur and calculate the internal resistance of the cell.

Solution   When the key is open, no current flows through r or R at balance and so the p.d. across the wire is equal to the e.m.f. of the cell. As soon as the key is closed, current flows and a p.d. appears across the internal resistance of the cell. Therefore the new balance point occurs when the p.d. across the wire is $V = E - Ir$, where I is the current flowing. As $(E - Ir)$ is less than E, the new balance point, which is proportional to the p.d., is reduced. r can be found from equation (14:13), though it must be stressed that an understanding of how this result is obtained is better than merely remembering the formula.

$$r = R\left(\frac{l}{l'} - 1\right) = 5\left(\frac{78.1}{50.1} - 1\right) = 2.8\ \Omega$$

14:13   Measurement of Resistance

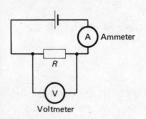

Fig. 14:19 Voltmeter and ammeter used to determine resistance of a conductor.

Fig. 14:19 shows an arrangement consisting of an ammeter and a voltmeter which could be used to determine the resistance R of the conductor. If I is the current registered on the ammeter, and V is the p.d. registered on the voltmeter, then $R = V/I$. However the voltmeter and ammeter affect the quantities they are trying to measure, and so this method is not suitable for an *accurate* determination of R.

The **Wheatstone bridge** does offer an accurate method of measuring resistances. Fig. 14:20 shows the Wheatstone bridge arrangement. R_A, R_B and R_C are resistors of known resistance and R_U is the resistance of unknown value. The driving cell is used to drive current through all four resistors and the galvanometer is used to detect current flow between S and T. The bridge is said to be "balanced" when no current flows between S and T, this being achieved by adjusting the value of one of the known resistors. By considering the conditions for balance to occur, we are able to arrive at a formula for the value of the unknown resistance.

At point P the current divides; I_1 travels through resistor R_A and then through resistor R_C (there is no division of current at T because at balance *no* current passes through the galvanometer). Similarly current I_2 travels through resistor R_B, then resistor R_U. At balance, the p.d. between S and T is zero so that points S and T are at the same potential. It follows that the p.d. across resistor R_A must equal the p.d. across R_B or

$$V_{PS} = V_{PT} \tag{14:14}$$

Similarly the p.d. across R_C must equal p.d. across R_U or

$$V_{SQ} = V_{TQ} \tag{14:15}$$

Now $V_{PS} = I_1 R_A$ $V_{PT} = I_2 R_B$ $V_{SQ} = I_1 R_C$ $V_{TQ} = I_2 R_U$

Substituting these into equations (14:14) and (14:15) and dividing (14:15) by (14:14) gives

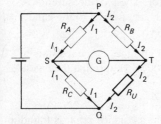

Fig. 14:20 The Wheatstone Bridge.

163

$$\frac{I_1 R_C}{I_1 R_A} = \frac{I_2 R_U}{I_2 R_B}$$

It follows that

$$R_U = \frac{R_C R_B}{R_A} \tag{14:16}$$

Hence if R_A, R_B and R_C are known, R_U can be determined.

Exercises

14.1 If a current of 2 A flows round an electrical circuit, how much charge passes a point in 3 min?

14.2 State Ohm's law. If the p.d. across a conductor is 5 V and the current flowing through the conductor is 5 A, calculate the resistance of the conductor.

14.3 A copper wire of radius r and length l has resistance R. Determine the resistance of another copper wire of radius $r/2$ and length $2l$, in terms of R.

14.4 Two wires of the same length have resistances in the ratio $3:1$. If their radii are in the ratio $1:2$, calculate the ratio of their resistivities.

14.5 The p.d. between the terminals of a battery is 2.04 V. Calculate how much work is done when a charge of 2 C moves between the terminals. If it takes 3 seconds for the charge of 2 C to move between the terminals, calculate the power dissipated in the circuit.

14.6 A current of 4 A passes through a heater that has a resistance of 60 Ω. Calculate
   a) The p.d. across the heater.
   b) The power dissipated in the heater.
   c) The total energy converted in 1 hour.

14.7 The resistance of a specimen of copper is 5 Ω at 20°C. If the temperature coefficient of copper is $4 \times 10^{-3}/°C$, calculate the resistance of the specimen at 100°C.

14.8 Distinguish between the terms e.m.f. and potential difference.

14.9 An accumulator of e.m.f. 2.0 V and negligible internal resistance is connected to two resistors in series of resistance 4 Ω and 10 Ω. Calculate
   a) The current flowing through the circuit.
   b) The power dissipated in each resistor.
If the same two resistors are now connected in parallel, calculate
   c) The current flowing through each resistor.
   d) The power dissipated in each resistor.

14.10 A dry cell of e.m.f. 1.5 V is connected to a 3 Ω resistor. The p.d. measured across the terminals of the cell is 1.0 V. Explain why the p.d. is not equal to the e.m.f. of the cell and calculate the internal resistance of the cell.

14.11 Describe the principle of operation of the Wheatstone bridge. Three resistors in a balanced bridge have resistances 500 Ω, 300 Ω and 700 Ω. Determine the possible values of the other resistor.

14.12 A potentiometer wire of length 2 m has a p.d. between its ends of 2.2 V. A cell of unknown e.m.f. is connected in the usual manner and a balance point is found a distance of 138 cm from one end of the potentiometer wire. Calculate the e.m.f. of the cell.

15 Moving Charges and Magnetism

Chapter 13 was concerned with the forces between stationary charges. When charges are moving there is another force that has to be accounted for which is quite different to the electrostatic interaction. This is called the magnetic interaction and in this chapter we will consider magnetic effects, outlining some of their more important applications.

15:1 Magnetism

Fig. 15:1 Small bar magnet pivoted freely in the earth's magnetic field.

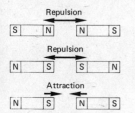

Fig. 15:2 Summary of forces experienced by two bar magnets in close proximity.

Steel amongst other materials can exhibit magnetic effects. Because the magnetic effects produced by certain magnetic materials do not decrease with time, they are called **permanent magnets** and are used in such devices as the electric motor, galvanometer and the dynamo (which will be dealt with in sections 15:11, 15:12 and 16:10 respectively). Magnetic effects, however, are not restricted to permanent magnets. The Earth for example has its own magnetic field (see section 15:2) and it was because permanent magnets aligned themselves in this field that the ends of the magnets became known as North and South poles. Electric currents also display magnetic effects and it is from the study of moving charges that a full explanation of all the effects mentioned briefly here can be understood.

A small bar magnet shown in fig. 15:1 pivoted about its centre of gravity normally aligns itself to point in a North–South direction. The end of the magnet that points North is called the **North pole** and the end that points South, the **South pole**. Fig. 15:2 shows the direction of the forces experienced by two magnets brought close together. It is found that if two North poles or two South poles are brought close together they will repel, whereas a North pole and a South pole close together will attract. This can be summarised as follows,

Like poles repel but unlike poles attract.

Note the analogy with charges: like charges repel but unlike charges attract. It is important to stress, however, that magnets carry *no* net charge so that forces of attraction or repulsion are *not* due to electrostatic charges.

15:2 The Magnetic Field and Magnetic Field Lines

A magnet has the ability to exert a force on another magnet placed close to it. There exists in the region surrounding a magnet a **magnetic field**. Any magnet placed in a magnetic field experiences a force due to that field. Magnetic field lines are drawn as an aid to visualising a magnetic field. There is an arrow on each line which indicates the direction of the force that a North pole of a magnet would experience if placed in the magnetic field.

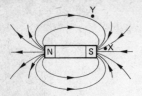

Fig. 15:3 Magnetic field lines in vicinity of a bar magnet.

The direction of a magnetic field at a point in space can be investigated experimentally using a plotting compass. This is simply a very small bar magnet pivoted freely so that it can respond to a magnetic field by aligning itself with the magnetic field lines. The magnetic field associated with a bar magnet is shown in fig. 15:3. This can be verified by placing a plotting compass at any point around the bar magnet and noting which way the North pole of the compass points. It appears that the magnetic field lines originate at the North pole and terminate at the South pole. The number of field lines per unit area indicates the strength of the magnetic field so that, for example, the magnetic field shown in fig. 15:3 is stronger at point X than point Y owing to the fact that the density of field lines at X is greater than at Y.

Fig. 15:4 shows the field line configuration for two North poles, two South poles, and a North and South pole in close proximity. It can be seen that a number of field lines appear to stop abruptly in space but in actual fact every field line loops back onto itself and it is the limitation of space on the page that prevents every complete loop from being drawn.

Fig. 15:4 Magnetic field lines in the vicinity of two bar magnets: (a) when a North and a South pole are close together; (b) when two North poles are close together; (c) when two South poles are close together.

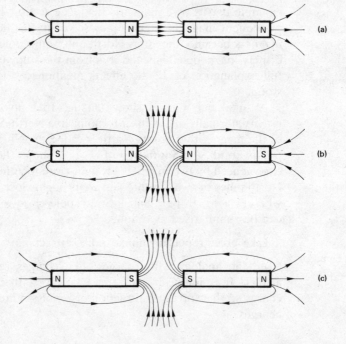

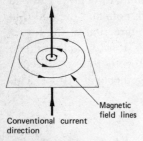

Conventional current direction

Magnetic field lines

Fig. 15:5 Magnetic field lines surrounding current-carrying conductor in a plane perpendicular to the direction of the current.

15:3 Magnetic Field Surrounding a Straight Conductor Carrying a Current

A magnetic field is caused by moving charges. This can be verified experimentally by holding a plotting compass close to a current-carrying wire. Fig. 15:5 shows the magnetic field lines surrounding a long wire carrying a current. The magnetic field lines form concentric circles centred upon the wire. The direction of the magnetic field depends on the direction of the current and a convenient aid to remembering the relation between the current and field direction is Maxwell's corkscrew rule.

Fig. 15:6 Maxwell's cork-screw rule.

(a)

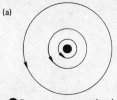

● Represents a conventional current emerging perpendicularly OUT FROM the plane of the page

(b)

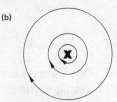

✗ Represents a conventional current going INTO the page perpendicular to the plane of the page.

Fig. 15:7 Direction of magnetic field lines associated with (a) conventional current coming out of paper; (b) conventional current going into paper.

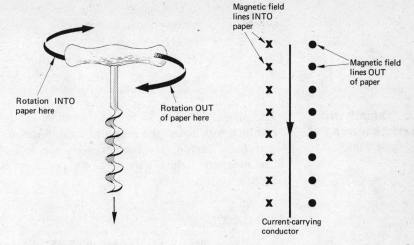

Imagine rotating a corkscrew into a block of wood. If the direction of the forward motion of the point of the corkscrew is taken to be the direction of conventional current flow, then the direction of rotation of the corkscrew handle is equivalent to the direction of the magnetic field lines (see fig. 15:6).

Fig. 15:7 shows a conventional current flowing into and out of the paper with the associated magnetic field found using Maxwell's corkscrew rule. The magnetic field lines are closest together near to the wire and gradually spread out as the distance from the wire increases, indicating that the magnetic field is strongest close to the wire.

15:4 Force between Current-carrying Conductors

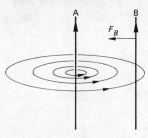

Fig. 15:8 Force on a current-carrying wire B in a magnetic field produced by a current flowing in wire A.

Fig. 15:9 Fleming's left-hand rule.

A magnetic field produced when charges move can exert a force on other moving charges. If two wires are placed side by side and a current flows through each wire in the same direction as shown in fig. 15:8, a force acts on each wire in such a direction that it tends to draw the two wires together. The magnetic field produced by the current in wire A exerts a force on wire B. Similarly the magnetic field produced by the current in wire B exerts a force on wire A. The direction of the force on wires A and B can be found using Fleming's left-hand rule.

Refer to fig. 15:9. Place the thumb and first two fingers of the left hand so that they are mutually perpendicular. If the forefinger points in the

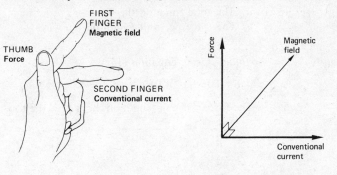

direction of the magnetic field, and the second finger points in the direction of conventional current flow, then the thumb points in the direction of the force on the wire.

15:5 Quantitative Description of a Magnetic Field

The magnetic field is defined in terms of the force on a current-carrying conductor placed in the magnetic field. Suppose a straight wire of length l carrying a current I experiences a force F in a uniform magnetic field. The **magnetic field** B existing at the position where the wire is situated is given by

$$B = \frac{F}{Il} \qquad (15:1)$$

The unit of magnetic field is the TESLA (symbol T). This formula can be rearranged to give the force on a current I flowing in a straight wire of length l in a uniform magnetic field B:

$$F = BIl$$

This formula holds true if the magnetic field and the current are perpendicular to each other. However the full formula for the force is

$$F = BIl \sin \theta \qquad (15:2)$$

where θ is the angle between the magnetic field and the conventional current direction. Referring to fig. 15:10, if $\theta = 90°$ then $F = BIl$ since $\sin 90° = 1$. If $\theta = 0°$, then $F = 0$ since $\sin 0° = 0$.

It can be seen therefore that a current can flow in the region of a magnetic field without experiencing a force so long as the current runs parallel to the magnetic field lines.

As the magnetic field has a magnitude (given by equation 15:1) and a direction (given by the arrows on the magnetic field lines), it is a vector quantity.

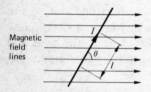

Magnetic field lines

Fig. 15:10 Force on a current carrying wire at an angle θ to magnetic field lines. The force on the wire is perpendicular, into the paper (Fleming's left-hand rule).

Example 15:1

A wire 2 cm long carrying a current of 10 A is placed perpendicular to a uniform magnetic field of 0.2 T. Calculate
  a) The total force on the wire.
  b) The force per unit length on the wire.

Solution   Using equation (15:1),

$$F = BIl = 0.2 \times 10 \times 2 \times 10^{-2} = 4 \times 10^{-2} \, \text{N}$$

The force per unit length on the wire is the total force divided by the length of the wire:

$$\frac{F}{l} = \frac{4 \times 10^{-2}}{2 \times 10^{-2}} = 2 \, \text{N/m}$$

15:6 Force on a Moving Charge in a Magnetic Field

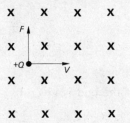

Fig. 15:11 Force on a charge Q moving in a magnetic field.

Fig. 15:11 shows a positive charge moving perpendicularly to a magnetic field. The direction of the force on the charge can be determined using Fleming's left-hand rule. This force is perpendicular to the direction of motion of the charge and has the effect of changing the direction of motion of the charge without altering its speed. The force experienced by the charge is a centripetal force (see section 2:7) and the path of the moving charge in the uniform magnetic field is circular.

It is possible to derive an expression for the **force on a charge** Q moving with a velocity v perpendicular to a uniform magnetic field B equivalent to the expression $F = BIl$. The equation is

$$F = BQv \qquad (15:3)$$

As stated this is a centripetal force and so we can equate equation (15:3) to the equation derived for the centipetal force acting on a body, mass m, moving with velocity v in a circle of radius r given in section 2:7, to give

$$BQv = \frac{mv^2}{r} \qquad (15:4)$$

15:7 Magnetic Field at a Point in Space due to Specific Current Geometries

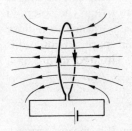

Fig. 15:12 Magnetic field lines associated with a loop of wire carrying a current.

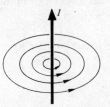

Fig. 15:13 Magnetic field lines in the vicinity of a long straight wire carrying a current.

The magnetic field at a point in space due to a current-carrying conductor depends on the magnitude of the current, the geometry of the conductor, and the medium in which the conductor is placed. The magnetic fields due to certain current geometries are given below without proof.

1 Fig. 15:12 shows the magnetic field lines associated with a **loop of wire** carrying a current. If the radius of the loop is r and the current carried by the loop is I, then the magnetic field at the centre of the loop is given by

$$B = \frac{\mu_0 I}{2r} \qquad (15:5)$$

where μ_0 is a constant known as the *permeability* of free space (see section 15:8).

2 Fig. 15:13 shows the magnetic field lines associated with a **long straight wire** carrying a current I. The magnetic field a distance r from the long wire is given by

$$B = \frac{\mu_0 I}{2\pi r} \qquad (15:6)$$

3 Fig. 15:14 shows the magnetic field lines associated with a **long solenoid** carrying a current I (note its similarity to the magnetic field produced by the bar magnet). If the solenoid has n turns per metre, then the magnetic field on the axis of the solenoid at the centre is given by

$$B = \mu_0 n I \qquad (15:7)$$

169

Fig. 15:14 Magnetic field line configuration produced by a solenoid carrying a current.

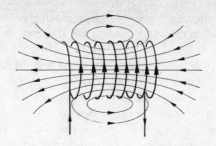

The magnetic field on the axis of the solenoid at the end is given by

$$B = \frac{\mu_0 n I}{2} \tag{15:8}$$

15:8 Permeability

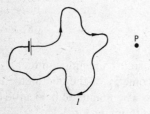

Fig. 15:15 Construction of magnetic field at P due to an arbitrary current configuration.

The magnetic field at point P shown in fig. 15:15 depends upon the charge flowing (the current), the path the charged particles take (which is determined by the configuration of the conductor), the distance from the conductor to the point P, and the medium in which the field is set up. The factor introduced that takes into account the medium is called the **permeability** of the medium. For a field in free space (a vacuum), the value of the permeability μ_0 is $4\pi \times 10^{-7}$ H/m (henrys per metre). [The henry is the unit of inductance which is considered in greater detail in the next chapter.] Certain materials such as iron and nickel have permeabilities that are many times greater than that of free space. If such materials are placed inside a current-carrying solenoid, for example, magnetic fields can be produced that are much greater than the solenoid is capable of producing in free space. The permeability of air, to a close approximation, is the same as the permeability of free space, so equations (15:5) to (15:8) apply to currents flowing in wires situated in air or a vacuum.

Example 15:2

Two long straight wires are placed 5 cm apart in a vacuum. If each wire carries a current of 10 A

    a) Calculate the force per unit length on each wire.

    b) Indicate the direction of the force on each wire if the currents run (i) in the same direction in each wire, (ii) in opposite directions in each wire.

Solution

a) Initially calculate the magnetic field in the region of wire B due to the current flowing through wire A. This is given by equation (15:6):

$$B_\text{A} = \frac{\mu_0 I}{2\pi r} = \frac{4\pi \times 10^{-7} \times 10}{2\pi \times 0.05} = 4 \times 10^{-5}\,\text{T}$$

To calculate the force on wire B we use equation (15:1):

$$F_\text{B} = B_\text{A} I l$$

Force/unit length $\dfrac{F_\text{B}}{l} = B_\text{A} I = 4 \times 10^{-5} \times 10 = 4 \times 10^{-4}\,\text{N/m}$

b) The force on wire B tends to pull it towards wire A (this can be found using Fleming's left-hand rule). If the force on wire A due to the magnetic field produced by the current in wire B is calculated, the magnitude of the force is found to be the same as that on wire B. Again using Fleming's left-hand rule, the force on wire A is found to be directed towards wire B.

If the direction of the current in one of the wires is reversed, the magnitude of the force per unit length is found to be the same as that calculated in part *a*). However instead of the force on each wire acting to pull the wires together, the forces tend to push the wires apart, i.e. there is mutual repulsion.

15:9 Definition of the Ampere

The force between two current-carrying conductors forms the basis of the definition of the ampere, to which other electrical quantities such as charge can be related. The unit of current, the ampere, is directly related to the fundamental quantities of force and length:

The **ampere** is the current which, when flowing through two straight, parallel conductors of negligible cross-section that are infinite in length, held one metre apart in a vacuum, causes a force of 2×10^{-7} N/m to be experienced by each conductor.

It is impracticable to use two infinite parallel conductors to measure current but fig. 15:16 shows a possible method of measuring a current in terms of the force between two current-carrying conductors. The torque (see section 3:1) produced by the two coils can be shown to be $T = kI^2$, where k is a constant. This anticlockwise torque can be balanced by a

Fig. 15:16 The current balance.

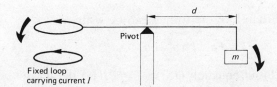

Fixed loop carrying current I

clockwise torque resulting from the force on the mass m at the other end of the balance. This torque is given by $T = mgd$, where d is the distance from the pivot point to end of the balance. Equating the two torques gives

$$kI^2 = mgd \qquad (g \text{ is the acceleration due to gravity})$$

m, g, d can be measured accurately and k can be evaluated from the geometry of the conductors. Therefore I can be determined by rearranging the above equation:

$$I = \sqrt{\frac{mgd}{k}} \qquad\qquad (15:9)$$

Note that, as this method involves only the measurement of mechanical quantities, it is termed an *absolute* measurement of current and the device outlined above is called a *current balance*.

15:10 Torque on a Coil in a Magnetic Field

As a wire carrying a current in a magnetic field experiences a force, it should be possible to make that force do work. A device that is designed to do just this is the electric motor. The first stage in understanding the electric motor is to investigate the torque produced on a coil carrying a current that is placed in a magnetic field. Consider the "square" coil shown in fig. 15:17(a). By using Fleming's left-hand rule, the direction of the force on each side of the square loop can be determined.

The forces on side PQ and RS are upward and downward respectively and as such are trying to stretch the coil. If the coil is rotated through 180°, the forces on these sides act to compress the coil. It can be seen therefore that the forces on PQ and RS do *not* produce a torque and can be neglected.

However the forces on SP and QR *do* produce a torque and, as the coil is pivoted freely about the axis AB, those forces tend to turn the coil in a clockwise direction. The torque on the coil is given by the product of the force and the perpendicular distance to the pivot. If the distance between P and Q is $2a$ and the axis of rotation AB is situated equidistant from P and Q, then the torque acting about this axis, when the plane of the coil is parallel to the field, is

$$T = F_{SP}a + F_{QR}a$$

where F_{SP} and F_{QR} are the forces on sides SP and QR respectively. If the coil is in a uniform magnetic field then the magnitude of F_{SP} equals the magnitude of F_{QR} so we can write

$$F_{SP} = F_{QR} = F$$

Therefore the torque can be written

$$T = Fa + Fa = 2Fa$$

From equation (15:1), $F = BIl$ so the torque becomes

$$T = 2BIla$$

The area A bounded by the coil is $2al$, giving $T = BIA$. If there are N turns on the coil, then a torque of magnitude BIA acts on *each* turn of the coil. Therefore the **torque acting on the whole coil** is given by

$$T = BIAN \qquad (15:10)$$

This equation applies when the plane of the coil is parallel to the magnetic field lines. As rotation from this position occurs, then although the forces on the side of the coil remain constant, the perpendicular distance from the line of action of the force to the the pivot decreases to zero when the plane of the coil is perpendicular to the magnetic field lines, and so the torque decreases to zero also. The torque on the coil will cause it to rotate to a position where the plane of the coil is perpendicular to the magnetic field lines. The angular momentum of the coil may make it rotate past this position to that given in fig. 15:17(b). By using

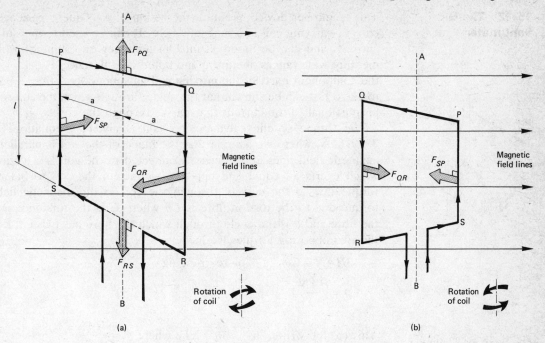

(a) (b)

Fig. 15:17 Torque on a coil in magnetic field.

Fleming's left-hand rule it can be seen that the torque now acting on the coil will make it rotate in an anticlockwise direction; that is a full rotation of 360° will not occur, and the coil will eventually come to rest with its plane perpendicular to the magnetic field.

In order to lift or move objects, constant rotation in one direction only is required and this is accomplished in the d.c. motor.

15:11 The Direct Current (d.c.) Motor

Fig. 15:18 shows a simple **d.c. motor**. Its construction is very similar to the coil in the magnetic field described in the previous section. The magnetic field in the region of the coil is usually provided by a permanent magnet in the case of small motors. The extra component is a *commutator* which consists of two semicircular strips of metal separated by an insulator.

When a current flows through the coil it rotates in an anticlockwise direction. When the coil is perpendicular to the magnetic field lines, the momentum gathered by the coil will be sufficient to turn the coil so that section Y of the commutator comes into contact with brush X and section Z of the commutator comes into contact with brush W. As this happens, current flows in at section Z of the commutator and out of section Y. By using Fleming's left-hand rule we see that the direction of the torque on the coil is unchanged and the rotation of the coil continues in an anticlockwise direction.

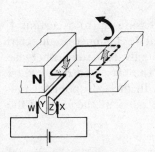

Fig. 15:18 Simple d.c. motor. The coil rotates in an anticlockwise direction.

The coil will continue to rotate as long as the current is maintained in the coil. The electrical energy supplied to the coil is converted into mechanical energy by the action of the motor.

15:12 The Galvanometer

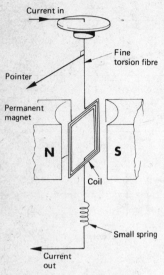

Current in

Fine torsion fibre

Pointer

Permanent magnet

N   S

Coil

Small spring

Current out

Fig. 15:19 Schematic of a moving coil galvanometer.

This is another device based on the action of a torque experienced by a current-carrying coil in a magnetic field. It can be used to measure small currents and can be modified into an ammeter or voltmeter which can measure wide ranges of current and potential difference. Fig. 15:19 shows the component parts of the **moving coil galvanometer** (often abbreviated to galvo.). It can be shown that the angle through which the coil rotates is proportional to the current that passes through the coil.

The torque on the galvanometer coil is given by equation (15:10), $T = BIAN$, where we assume that the plane of the coil is parallel to the magnetic field lines at all times.† Connected to the coil is a torsion fibre which exerts an equal and opposite torque on the coil to balance the torque due to the current through the coil in the magnetic field. The torque due to the torsion fibre is $C\theta$ where C is the torsion constant of the fibre and θ is the angle through which the fibre has twisted. Equating the two opposing torques we have

$$BIAN = C\theta \qquad \text{or rearranging}$$

$$\theta = \frac{BAN}{C}I \qquad\qquad\qquad (15:11)$$

This can be written as $\qquad \theta = KI \quad$ where $K = \dfrac{BAN}{C}$

We can see from equation (15.11) that the angle through which a pointer attached to the coil will turn, is directly proportional to the current flowing through the coil. To increase the sensitivity of the galvanometer, the constant K must be increased. This can be done in principle by increasing B, A, and N or decreasing C.

15:13 Modification of the Galvanometer to Act as an Ammeter

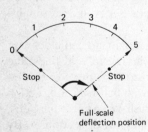

1 2 3 4
0   5
Stop   Stop

Full-scale deflection position

Fig. 15:20 Full-scale deflection of a meter.

Suppose a galvanometer has a full-scale deflection which occurs when the current through the coil is I_m. Full-scale deflection (often abbreviated to f.s.d.) refers to the maximum deflection of the pointer which indicates the maximum current, as shown in fig. 15:20. If the resistance of the galvanometer coil is R_c then the p.d. across the terminals of the galvanometer when f.s.d. occurs is given by

$$V_{fsd} = I_m R_c$$

If this galvanometer is to be modified to measure up to a current I, where I is greater than I_m, the "excess" current $I - I_m$ must be directed elsewhere, leaving only I_m to flow through the galvanometer; otherwise the galvanometer could be seriously damaged. The excess current can be directed through a resistor placed in parallel with the galvanometer coil (often referred to as a *shunt resistor*). In order to calculate the value of the shunt resistance R_s, consider a specific example.

† *Note* The coil normally surrounds a fixed soft iron cylinder. This increases the magnetic field and ensures that the magnetic field lines are perpendicular to the coil regardless of the position of the coil.

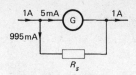

Fig. 15:21 Galvanometer modified to act as an ammeter.

A galvanometer has to be modified to measure currents up to 1 A. F.s.d. occurs when the current through the galvanometer coil is 5 mA and the resistance of the coil is 20 Ω.

$$V_{fsd} = I_m R_c = 5 \times 10^{-3} \times 20 = 100 \text{ mV}$$

Refer to fig. 15:21. At f.s.d. 5 mA must pass through the coil of the galvanometer; it follows that 1 A − 5 mA = 995 mA must be directed through the shunt resistance R_s. As the galvanometer and the shunt resistor are in parallel, the p.d. across each must be the same, that is 100 mV. R_s can now be found from

$$R_s = \frac{V_{fsd}}{I_s} = \frac{100 \times 10^{-3}}{995 \times 10^{-3}} = 0.1005 \text{ } \Omega$$

15:14 Modification of the Galvanometer to Act as a Voltmeter

An instrument used to measure p.d.s is called a voltmeter. It is possible to modify the galvanometer to measure a range of p.d.s. If a p.d. of greater than V_{fsd} appears across the galvanometer coil (see section 15:13) the pointer will deflect "off-scale" and the meter may be damaged. Therefore if a p.d. of up to V volts is to be measured, the excess p.d., $V - V_{fsd}$, must be accommodated elsewhere. This is done by placing a resistor in *series* with the galvanometer coil as shown in fig 15:22. Consider a specific example.

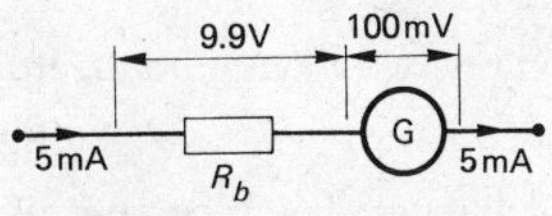

A voltmeter is required that will measure a p.d. of up to 10 V. The resistance of the coil is 20 Ω and f.s.d. occurs when a current of 5 mA passes through the galvanometer coil. When f.s.d. occurs there is a p.d. of 100 mV across the galvanometer coil. Therefore the "excess p.d." 10 V − 100 mV = 9.9 V must appear across the resistor R_b placed in series (sometimes referred to as a *bobbin resistor*). The current through the galvanometer at f.s.d. is 5 mA, and as R_b is in series with the galvanometer coil, 5 mA must also pass through it. It follows that

$$R_b = \frac{V_b}{I_m} = \frac{9.9}{5 \times 10^{-3}} = 1980 \text{ } \Omega$$

where V_b is the p.d. across the resistor R_b and I_m is the current through the coil at f.s.d.

Fig. 15:22 Galvanometer modified to act as a voltmeter.

Example 15:3

A galvanometer has a coil of resistance 10 Ω and f.s.d occurs when the current through the coil is 5 mA. Determine how the galvanometer could be modified into
  a) an ammeter that will measure up to 10 A
  b) a voltmeter that will measure up to 20 V

Solution   Refer to fig. 15.23.
a) p.d. across galvo at f.s.d. = $IR_s = 5 \times 10^{-3} \times 10 = 50$ mV
A shunt resistor placed in parallel with the galvo must take a current of 10 A − 5 mA = 9.995 A. It follows that

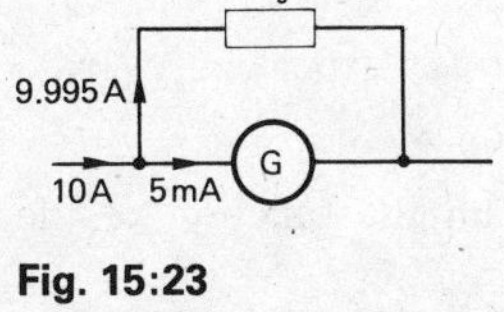

$$R_s = \frac{50 \times 10^{-3}}{9.995} = 5.0025 \times 10^{-3} \text{ } \Omega$$

Fig. 15:23

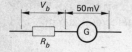

Fig. 15:24

b) Refer to fig. 15:24

The p.d. across R_b is $20\,\text{V} - 50\,\text{mV} = 19.95\,\text{V}$ and, as the current through both R_b and the galvanometer coil is the same (5 mA),

$$R_b = \frac{19.95}{5 \times 10^{-3}} = 3.99 \times 10^3\,\Omega$$

Exercises

15.1 Explain the cause of a magnetic field and the condition necessary for a current-carrying conductor in a magnetic field to experience no force.

15.2 A wire of length 0.5 cm carries a current of 2 A. When the wire is placed perpendicular to a magnetic field it experiences a force of $3 \times 10^{-2}\,\text{N}$. Calculate the value of the magnetic field.

15.3 Draw the magnetic field line configuration when a North pole and a South pole of two bar magnets are in close proximity.

15.4 An electron (charge $1.6 \times 10^{-19}\,\text{C}$ and mass $9.1 \times 10^{-31}\,\text{kg}$) travels at a velocity of $2 \times 10^7\,\text{m/s}$ perpendicular to a uniform field of 0.1 T. Calculate the radius of the path of the electron.

15.5 A long solenoid has 50 turns/cm and carries a current of 2 A. Calculate the magnetic field at the end of the solenoid. If a small wire of length 2 mm carrying a current of 3 A is placed perpendicular to the magnetic field lines at the end of the solenoid, calculate the force experienced by the wire.

15.6 Calculate the magnetic field a distance 20 cm from a long wire carrying a current of 2 A.

15.7 Explain the operation of the current balance. A current balance has a current of 3 A flowing through its coils and is in equilibrium when a mass of 3 g is suspended from a point 30 cm from the pivot. If the current is increased to 4 A determine what additional mass must be added to maintain equilibrium.

15.8 A coil of area $10\,\text{cm}^2$ has its plane parallel to a uniform magnetic field of 0.1 T. If the coil has 100 turns and a current of 1 A flows through it, calculate the torque experienced by the coil.

15.9 Calculate the magnetic field at the centre of a solenoid 2 m long, which has 2000 turns and carries a current of 3 A.

15.10 State the units of a) magnetic field, b) torque, c) current.

15.11 Explain the operation of the galvanometer and show that the angle of rotation of the galvanometer coil is proportional to the current through the coil. A galvanometer has a coil of resistance 15 Ω and f.s.d. occurs when the current through the coil is 10 mA. Explain how the galvanometer may be modified into
 a) An ammeter capable of reading up to 30 A.
 b) A voltmeter capable of reading up to 100 V.
In both a) and b) state the effective resistance of the instrument produced.

15.12 List factors which determine the magnitude of the torque experienced by a current-carrying coil in a uniform magnetic field.

15.13 Describe the construction of a simple d.c. motor. Explain the function of the commutator.

16 Electromagnetic Induction

16:1 Effect of Moving a Wire through a Magnetic Field

If the wire shown in fig. 16:1 is moved upwards across a magnetic field, then the electrons within the wire also move upwards. The motion of the electrons upwards is equivalent to a flow of conventional current downwards. Fleming's left-hand rule indicates that the force on the electrons is such that they move towards end M of the wire, thus making that end negative. Owing to the fact that the wire is overall electrically neutral, end N must become positively charged. The consequence of this is that an electromotive force is set up between M and N. This e.m.f. will continue for as long as the force remains on the electrons, and this occurs provided the wire continues to move in the magnetic field. When the wire is brought to rest, the e.m.f. between the ends of the wire decreases to zero.

Fig. 16:1 Direction of conventional current produced when a wire moves through a magnetic field.

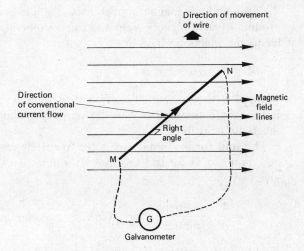

If a circuit is completed between the ends M and N, then a current can be made to flow around the circuit so long as the wire moves through the magnetic field. By moving the wire alternately upwards and downwards through the magnetic field, the pointer on the galvanometer is made to swing first one way and then the other, indicating that the direction of motion of the wire is one factor that determines the direction of the current through the circuit. It can be shown experimentally that the magnitude of the e.m.f. set up between the ends of the wire depends on the strength of the magnetic field, the speed at which the wire moves through the magnetic field, the length of the wire in the magnetic field, and the angle between the direction of motion of the wire and the direction of the magnetic field lines.

16:2 Magnetic Flux

Before quantitatively describing the e.m.f. produced by the motion of a conductor in a magnetic field, a quantity called the magnetic flux must be defined. Fig. 16:2 shows a uniform magnetic field B passing through an

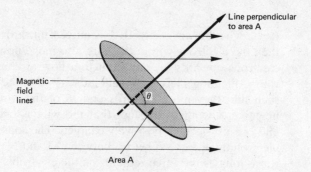

Fig. 16:2 Magnetic flux related to magnetic field.

area A. θ is the angle between a line drawn perpendicular to the plane of the area and the magnetic field lines. The **magnetic flux** is defined as the product of the magnetic field and the component of the area perpendicular to the magnetic field lines:

$$\Phi = BA \cos \theta \qquad\qquad (16:1)$$

where Φ is the magnetic flux, B is the magnetic field, A is the area, and θ is the angle between a line drawn perpendicular to the area and the magnetic field lines.

The unit for magnetic flux is the WEBER (symbol Wb). By arranging equation (16:1) we can see that an alternative unit for magnetic field is given by

$$B = \frac{\Phi}{A} = \frac{\text{webers}}{(\text{metres})^2} \quad \text{or} \quad \text{Wb/m}^2 \quad (\text{assuming } \cos \theta = 1)$$

It follows that $1 \text{ Wb/m}^2 = 1 \text{ T}$. As the magnetic field is the magnetic flux per unit area, it is otherwise known as the *magnetic flux density*.

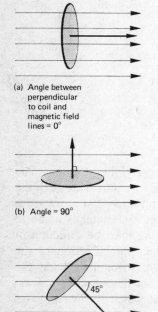

(a) Angle between perpendicular to coil and magnetic field lines = 0°

(b) Angle = 90°

(c) Angle = 45°

Fig. 16:3

Example 16:1

A circular coil of radius 10 cm is orientated in a constant magnetic field of 0.1 T as shown in fig. 16:3. Determine the magnetic flux through each coil in the three cases.

Solution
a) $\Phi = BA \cos 0° = 0.1 \times \pi \times (0.1)^2 \times 1 = 3.14 \times 10^{-3} \text{ Wb}$ $(\cos 0° = 1)$
b) $\Phi = BA \cos 90° = 0.1 \times \pi \times (0.1)^2 \times 0 = 0$ $(\cos 90° = 0)$
c) $\Phi = BA \cos 45° = 0.1 \times \pi \times (0.1)^2 \times 0.707 = 2.22 \times 10^{-3} \text{ Wb}$
       $(\cos 45° = 0.707)$

16:3 Faraday's Law

Faraday's law states: when magnetic field lines are crossed by a conductor (sometimes referred to as "cutting the magnetic field lines"), an e.m.f. is *induced* between the ends of the conductor by virtue of the force on the free electrons as they are moved through the magnetic field.

Faraday quantitatively described this **induced e.m.f.** as follows:

$$E = -\frac{d\Phi}{dt} \tag{16:2}$$

(The significance of the minus sign will be dealt with later.)

$\dfrac{d\Phi}{dt}$ is termed the *rate of change of flux with respect to time.*

When the wire MN of length l shown in fig. 16:4 moves at a steady velocity v, the area it traces out per second is lv. The wire crosses the

Fig. 16:4 E.m.f. induced in wire moving in a magnetic field.

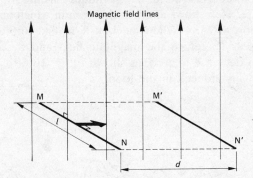

Magnetic field lines

magnetic field lines and the greater its velocity, the greater the number of magnetic field lines crossed per second. If the wire moves from MN to M'N' in a time t, then the number of magnetic field lines crossed per second is given by

$$\frac{d\Phi}{dt} = \frac{BA}{t}$$

Now $A = ld$ so replacing A gives

$$\frac{d\Phi}{dt} = \frac{Bld}{t} \quad \text{but} \quad \frac{d}{t} = v \quad \text{so} \quad \frac{d\Phi}{dt} = Blv$$

Using Faraday's law, the induced e.m.f. between the ends of the wire is given by

$$E = -\frac{d\Phi}{dt} = -Blv \tag{16:3}$$

Further application of Faraday's law can explain the e.m.f. induced in a stationary coil situated in a changing magnetic field. Fig. 16:5 shows a bar magnet moving towards a loop of wire. The magnetic flux that links† the

---

† If a uniform magnetic field B passes perpendicularly through a coil of area A, then the flux "linking" the coil is the total magnetic flux through the coil given by $\Phi = BA$. If the coil has N turns, then the flux linkage is given by $N\Phi$ since Φ is the flux linking each turn.

Fig. 16:5 Changing magnetic flux linking a loop of wire due to the motion of a magnet.

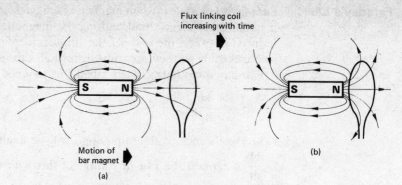

loop increases (because the magnetic field increases) and so by Faraday's law there must be an e.m.f. induced in the loop.

Fig. 16:6 shows another situation in which an e.m.f. may be induced in a loop of wire. When the key K in fig. 16:6(a) is opened, current flow ceases. Therefore the magnetic field and the magnetic flux linking the loop decrease to zero, as shown in fig. 16:6(b). As the flux decreases, an e.m.f. is induced in the loop.

Fig. 16:6 Changing magnetic flux linking a loop of wire due to changing current through a solenoid.

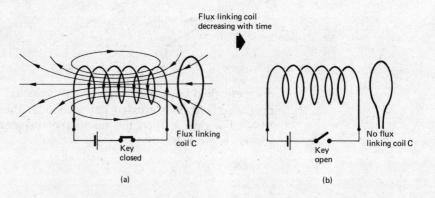

If an e.m.f. is induced in a loop or coil it can make a current flow if the ends of the loop or coil are connected to an external circuit. The direction of the current produced by the induced e.m.f. can be discovered using *Fleming's right-hand rule.*

16:4 Fleming's Right-hand Rule

Refer to fig. 16:7. Place the thumb and the first two fingers of the right hand so that they are mutually perpendicular. If the thumb points in the direction of motion of the wire and the forefinger points in the direction of the magnetic field, then the second finger points in the direction of the current.

Fig. 16:8 shows a wire moving through a magnetic field connected to an external circuit. Fleming's right-hand rule indicates that the direction of current flow is from A to B. Now a current that flows perpendicular to a magnetic field experiences a force and the direction of that force is given

Fig. 16:7 Fleming's right-hand rule.

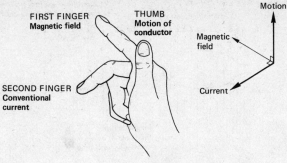

Fig. 16:8 Conventional current direction found using Fleming's right-hand rule.

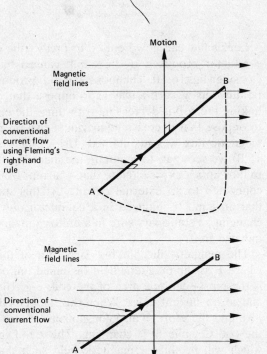

Fig. 16:9 Force on wire shown in fig. 16:8 found using Fleming's right-hand rule.

by Fleming's *left*-hand rule. Fig. 16:9 shows the direction of the force on the wire AB and it can be seen that this force is in the *opposite* direction to the motion that is responsible for the induced e.m.f. This "opposition to the change responsible for the e.m.f." is stated more generally in Lenz's law.

16:5   Lenz's Law

Lenz's law states: an induced e.m.f. acts in such a way as to oppose the change in magnetic flux that is producing it. The significance of the minus sign in equation (16:2) can now be understood. If, for example, the magnetic flux linking a coil decreases, then the induced e.m.f. in the coil will act to *oppose* that change and attempt to maintain the magnetic flux linking the coil. If on the other hand the flux linking a coil is increasing with time, then the induced e.m.f. will act to oppose this increase.

Fig. 16:10 Lenz's law used to find current in wire situated in a changing magnetic field.

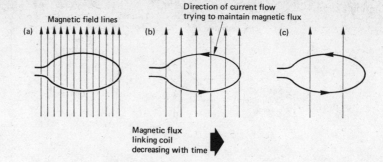

Lenz's law can now be used to predict the direction of the current in a conductor produced by an e.m.f. caused by a changing magnetic flux. Consider fig. 16:10. The induced e.m.f. produced by the changing magnetic flux acts in such a way as to oppose that change. If the magnetic flux linking the coil is decreasing with time, the e.m.f. induced would attempt to oppose that decrease by driving a current round the circuit so that the magnetic flux was maintained. Using Maxwell's corkscrew rule (section 15:3) we see that, to maintain the magnetic flux, the current would have to flow anticlockwise. (Note that a current would only flow if the coil were connected to an external circuit.) At this stage it is worth emphasising that an e.m.f. is induced in a conductor *only* while the magnetic flux is changing; a stationary wire in a uniform magnetic field will have *no* e.m.f. induced in it.

The magnetic flux linking each turn of the coil in fig. 16.6(a) is BA where B is the magnetic field (assumed uniform over the area A) and A is the cross-sectional area of the coil. Fig. 16:6(b) shows the key opened and so no current flows. When this happens the magnetic field B goes to zero and therefore Φ also goes to zero. There will be an e.m.f. induced in the coil C while Φ is changing. This e.m.f. gives rise to a current in C which will produce a magnetic field such that it opposes the change in magnetic flux that is producing it. We have the situation depicted in fig. 16:11; the direction of the current required to oppose this change in magnetic flux is given by Maxwell's corkscrew rule.

In fig. 16:12 the magnetic flux linking the coil is changed by moving the bar magnet while keeping the coil stationary. The consequence again is that the induced e.m.f. drives a current which produces a magnetic field in such a direction that it opposes the change in the magnetic flux which is producing it.

If the magnetic flux changes, then the e.m.f. induced in *each* turn is

$$E = -\frac{d\Phi}{dt}$$

But there are N turns on the coil which act like N sources of e.m.f. in series, i.e. the total e.m.f. between the ends of the coil would be

$$E = -N\frac{d\Phi}{dt} \tag{16:4}$$

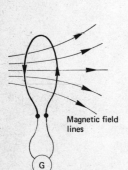

Fig. 16:11 Current direction opposing change in magnetic field depicted in fig. 16:6. Magnetic field through coil decreasing with time.

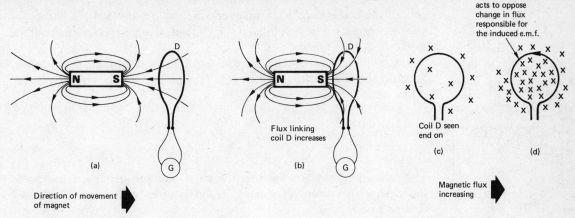

Current produced by induced e.m.f. acts to oppose change in flux responsible for the induced e.m.f.

Flux linking coil D increases

Coil D seen end on

(a)

(b)

(c)

(d)

Direction of movement of magnet

Magnetic flux increasing

Fig. 16:12 Current direction opposing change in magnetic field due to motion of a magnet.

Example 16:2

A coil of cross-sectional area $2.0 \times 10^{-2}\,\text{m}^2$ has 30 turns. The coil is situated perpendicular to a uniform magnetic field of 0.2 T. If the coil is removed from the magnetic field in 0.5 sec, determine the average e.m.f. induced between the ends of the coil.

Solution From equation (16:4) we have $\quad E = -N\dfrac{d\Phi}{dt}$

The average rate of change of flux can be written as $\Delta\Phi/\Delta t$ where $\Delta\Phi$ is the change of flux in a time interval Δt.

$$E = -N\frac{\Delta\Phi}{\Delta t} = -\frac{30 \times 0.2 \times 2.0 \times 10^{-2}}{0.5} = 0.24\,\text{V}$$

16:6 The Search Coil

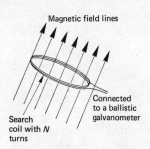

Magnetic field lines

Search coil with N turns

Connected to a ballistic galvanometer

Fig. 16:13 The search coil.

When a coil is withdrawn from a magnetic field, an e.m.f. is induced in the coil which is proportional to the magnetic field. Fig. 16:13 shows a **search coil** which is normally placed perpendicular to the magnetic field lines. As the search coil is quickly withdrawn from the magnetic field, an e.m.f. is induced in the coil which can cause a charge to flow in an external circuit. This charge can be detected by a ballistic galvanometer† and a full investigation shows that the magnetic field in the vicinity of the search coil is given by

$$B = \frac{kR}{NA}\,\theta$$

where R is the resistance of the circuit

N is the number of turns on the search coil

A is the area of the search coil and k is a constant

θ is the maximum angle of rotation of the galvanometer coil.

As B is proportional to θ, the search coil connected to a ballistic galvanometer can be used to measure magnetic field.

† The ballistic galvanometer is similar in construction to the galvanometer discussed in Chapter 15, but is modified so that the angle of rotation of the coil is proportional to the charge passed through the coil.

16:7 Self-inductance

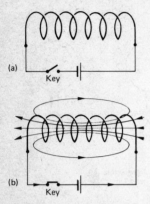

(a)

(b)

Fig. 16:14 Circuit used to explain self-inductance.

Consider fig. 16:14. When the key is closed, a constant current flows through the coil. It follows that a flux of BA links the coil, where A is the area of the coil and B is the magnetic field within the coil.† If the key is opened, the magnetic flux linking the coil will change and so there will be an e.m.f. induced in the coil.

It is found experimentally that the e.m.f. induced in the coil is proportional to the rate of change of current flowing through the coil. Expressed mathematically this becomes

$$E \propto -\frac{dI}{dt}$$

The minus sign appears because the e.m.f. induced acts so as to oppose the change in current producing it and is often called the *back e.m.f.* (Lenz's law again!)

The proportional sign can be replaced by a constant. This constant is called the **self-inductance** L of the coil. So we have

$$E = -L\frac{dI}{dt} \qquad (16:5)$$

The unit of inductance is the HENRY (symbol H) and the definition of the henry is as follows:

If an e.m.f. of one volt is induced in a conductor when the current flowing through the conductor changes at a rate of one ampere/sec, then the self-inductance of the conductor is one henry.

It can be shown that the self-inductance of a long solenoid of N turns, cross-sectional area A, and of length l can be given approximately by

$$L = \frac{\mu_0 N^2 A}{l} \qquad (16:6)$$

Equation (16:6) indicates that the self-inductance of a solenoid depends on the medium of the core, the number of turns on the solenoid, and the physical dimensions of the solenoid.

A coil or conductor of any configuration that is designed to have a large inductance is called an **inductor**. Any conductor has an inductance (just as any conductor also has a capacitance and a resistance). However, for a significant inductance, quantities such as N and A must be maximised. If iron is placed within the coil, then we can replace μ_0 by μ which can be a factor typically 1000 greater than μ_0, and so we have increased L by a factor of 1000. Placing a magnetic material with a large μ increases L and we can see that this is analogous to placing a dielectric of large ϵ in order to increase the capacitance of a parallel plate capacitor.

† Note that we have assumed B to be constant throughout the volume bounded by the coil; this is not in fact true but it does not affect the physical principles involved.

16:8 Mutual Inductance

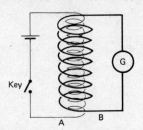

Fig. 16:15 Circuit used to explain mutual inductance.

Refer to fig. 16:15. Some of the magnetic field lines produced by the current-carrying solenoid A will pass through the solenoid B. While a steady current flows through the solenoid A, the magnetic flux within solenoid B will be constant. Suppose the key in solenoid A is opened so that the current ceases to flow. The magnetic flux within solenoid B must decrease to zero. A changing magnetic flux leads to an induced e.m.f. within solenoid B. The e.m.f. induced in B is proportional to the rate at which the current changes with time in solenoid A, i.e.

$$E_B \propto -\frac{dI_A}{dt}$$

where I_A is the current flowing through A. The minus sign appears again to show that the e.m.f. induced in solenoid B acts in such a way as to oppose the changing flux producing it. We can write

$$E_B = -M\frac{dI_A}{dt} \tag{16:7}$$

M is called the **mutual inductance** of the solenoids. The unit of mutual inductance is the same as that for self-inductance, namely the henry.

It can be shown that the mutual inductance of two long solenoids of N_A and N_B turns respectively, cross-sectional area A, in close proximity is given approximately by

$$M = \mu_0 \frac{N_A N_B A}{l} \tag{16:8}$$

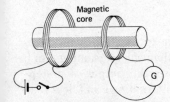

Fig. 16:16 Two arbitrary-shaped conductors.

The mutual inductance M depends on the dimensions of the conductor and the medium in which it is placed. It should be emphasised that this particular value of M is the approximate value for two long solenoids placed in close proximity. If the conductors were of some other shape (see fig. 16:16) then the mutual inductance would *not* be given by equation (16:8).

16:9 The Transformer

The transformer (shown in fig. 16:17(a)) uses the phenomenon of mutual inductance to "step up" or "step down" an alternating p.d.

An alternating source of e.m.f. has an e.m.f. versus time graph as shown in fig. 16:17(b) (also refer to chapter 17). As the e.m.f. connected to the primary coil in fig. 16:17(a) is continually changing, it follows that

Fig. 16:17 (a) The transformer; (b) alternating e.m.f. supplied to the primary coil of the transformer.

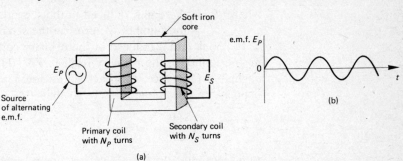

the current through the primary coil is also continually changing. Therefore the magnetic flux linking the primary and secondary coils changes with time and so an (alternating) induced e.m.f. appears between the ends of the secondary coil. The frequency of the induced e.m.f. is the same as that of the e.m.f. applied to the primary coil.

Fig. 16:18 Input e.m.f. E_P to primary coil and output e.m.f. E_S from secondary coil of a step-up transformer.

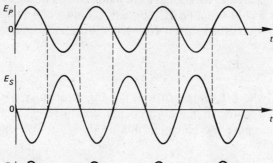

Fig. 16:19 Input e.m.f. E_P to primary coil and output e.m.f. E_S from secondary coil of a step-down transformer.

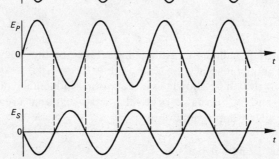

Fig. 16:18 shows the output of a **step-up transformer** which produces an output of alternating e.m.f. which is greater than the alternating e.m.f. placed across the primary coil. Fig. 16:19 shows the output of a **step-down transformer** which produces an output of alternating e.m.f. less than the alternating e.m.f. placed across the primary coil.

We assume that the magnetic flux which links the primary and secondary coils is the same. A laminated iron core is used to increase the mutual inductance of the coils (by increasing the magnetic flux linking the coils).

The induced e.m.f. in the secondary coil can be found using Faraday's law:

$$E_S = -N_S \frac{d\Phi}{dt}$$

where E_S is the e.m.f. induced in the secondary coil

N_S is the number of turns on the secondary coil.

The back e.m.f. induced in the primary coil as a result of the changing magnetic flux in the primary itself is $-N_P \, d\Phi/dt$. If the primary coil has negligible resistance, this back e.m.f. must, at any instant, be equal and opposite to the applied e.m.f., so that

$$E_P = N_P \frac{d\Phi}{dt}$$

Equating $d\Phi/dt$ in the above two equations gives

$$\frac{E_P}{E_S} = -\frac{N_P}{N_S} \qquad\qquad (16:9)$$

The minus sign arises because there is a π phase difference between the primary and secondary e.m.f.s as shown in figs. 6:18 and 6:19.

The conservation of energy principle (see section 4:4) tells us that it is impossible to get more energy out of the transformer via the secondary coil than we put into the primary coil. The best we can hope for is 100% transference of power from input to output, and in practice this can nearly be achieved.

Assuming there is no energy dissipation,

$$E_P I_P = E_S I_S$$

where I_P is the current in the primary coil and I_S is the current in the secondary coil. Therefore

$$\frac{E_P}{E_S} = \frac{I_S}{I_P}$$

Using (16:9) we get

$$\frac{E_P}{E_S} = \frac{I_S}{I_P} = \frac{N_P}{N_S} \qquad\qquad (16:10)$$

In practice, 100% transformer efficiency is never achieved. This is due to a number of factors:

1   The windings of the coil dissipate heat (the value of which will be $I^2 R$ where R is the resistance of the windings).
2   Heating losses within the transformer core due to currents set up in the core by the changing magnetic flux (known as *eddy currents*).
3   The transformer core is continually being magnetised and demagnetised by alternating currents flowing through the coils. Energy is used up during this process.
4   Not all the magnetic flux through the primary coil also links the secondary coil and so the e.m.f. induced in the secondary coil will not reach its theoretical maximum.

16:10   The Dynamo (or Generator)

We have considered a number of ways in which an e.m.f. can be induced in a conductor. Each time, the e.m.f. induced was due to a change of magnetic flux linking the conductor. If the motion of a conductor in a magnetic field can be maintained we will have a source of e.m.f. which could be a useful substitute for a cell or a battery. A device which does this is the dynamo. The **dynamo** at its most simple consists of a coil of N turns placed in a magnetic field as shown in fig. 16.20(a). If the magnetic field is uniform then the magnetic flux that links the coil when the plane of the coil is perpendicular to the magnetic field lines will be $N\Phi = NBA$, where A is the area of the coil. As the coil rotates about the axis AB, the flux linking the coil will change continuously from NBA to zero, when the plane of the coil is parallel to the magnetic field lines. A changing magnetic flux linking the coil results in an e.m.f. being induced in the coil.

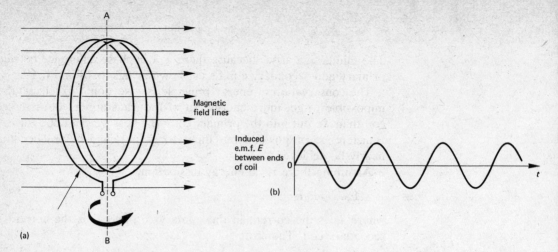

Fig. 16:20 (a) Rotating coil in a magnetic field; (b) output e.m.f. versus time for the dynamo.

Using Fleming's right-hand rule we can predict the direction of the induced e.m.f. from the movement of the coil through the field. The direction of the e.m.f. would change after every half-revolution of the coil. At a time t we have

$$\Phi = BA \cos \omega t$$

where ω is the angular velocity of the coil ($\omega = 2\pi f$). Now $E = -N \, d\Phi/dt$ so by differentiating the expression for Φ

$$\frac{d\Phi}{dt} = -BA\omega \sin \omega t$$

it follows that

$$E = N\omega BA \sin \omega t \qquad (16:11)$$

Fig. 16:20(b) shows a graph of e.m.f. versus time for the dynamo.

Example 16:3

A coil of cross-sectional area $10 \, \text{cm}^2$ is placed in a uniform magnetic field of 0.1 T. The coil has 250 turns.
    a) Calculate the magnetic flux linking the coil.
    b) If the coil rotates at 5 revolutions per sec calculate the maximum e.m.f. induced between the ends of the coil.

Solution

a) Magnetic flux linking the coil is $N\Phi$ where N is the number of turns on the coil.

$$N\Phi = 250 \times BA = 250 \times 0.1 \times 10 \times 10^{-4} = 2.5 \times 10^{-2} \, \text{Wb}$$

b) Refer to equation (16:11). The maximum e.m.f. occurs when $\sin \omega t = 1$; therefore

$$E = N\omega BA = 250 \times 2\pi \times 5 \times 0.1 \times 10 \times 10^{-4} = 0.785 \, \text{V}$$

Exercises

16.1 Describe a simple experiment to illustrate electromagnetic induction.

16.2 A wire of length 10 cm is moved at a velocity of 20 m/s perpendicular to a uniform magnetic field of 0.05 T. Calculate the e.m.f. induced between the ends of the wire. Explain why current does not flow in the wire.

16.3 A loop of wire of radius 1 cm is placed perpendicular to a magnetic field of 0.2 T. If the loop is rotated through an angle of 180° in 0.5 sec determine the average e.m.f. induced in the loop.

16.4 A solenoid 2 m long of radius 1 cm has 3000 turns. Give the approximate value of the self-inductance of the solenoid. Explain how this can be increased without altering the physical dimensions of the solenoid.

16.5 State Faraday's Law. Explain the significance of the minus sign.

16.6 Explain the principle of operation of the step-up transformer. A step-up transformer has 2000 turns on the primary coil and 5000 turns on the secondary coil. Assuming the transformer to be 100% efficient, calculate the secondary e.m.f. if an e.m.f. of 50 V is applied to the primary coil.

16.7 Describe the construction and operation of the dynamo.

16.8 A dynamo consisting of a coil of radius 3 cm has 200 turns and is rotated at a frequency of 50 revolutions per sec in a magnetic field of 0.01 T. Calculate the maximum e.m.f. output by the dynamo.

16.9 An inductive coil is connected in series to a battery having an e.m.f. of 1.5 V. Assuming the resistance is negligible and the current increases to 5 A in 3 sec determine the self-inductance of the coil.

16.10 Explain how the search coil may be used to determine the value of a magnetic field.

16.11 Give the main reasons for power loss in a transformer and explain what steps are taken to minimise this power loss.

17 Alternating Currents

17:1 Alternating E.M.F.

Chapters 14 and 15 dealt mainly with electric currents which do not vary with time. These are called *direct* currents. Such currents are produced by constant e.m.f.s, for example those of batteries or cells. To conclude the section on electricity and magnetism, alternating e.m.f.s, alternating potential differences and alternating currents will be briefly considered. In particular the variation of current and potential difference with time will be investigated for each of the circuit elements (resistor, capacitor and inductor) when a sinusoidally varying e.m.f. is applied to each in turn.

In the previous chapter it was seen that a rotating coil in a magnetic field would produce an **alternating e.m.f.** Referring to fig. 17.1, the maximum e.m.f. is called the **peak e.m.f.** The e.m.f. measured from the bottom of

Fig. 17:1 Definition of some terms used when dealing with alternating current.

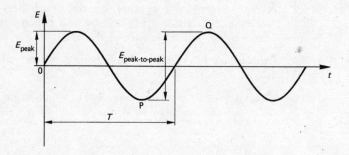

one trough (point P) to the top of the adjacent peak (point Q) is termed the **peak-to-peak e.m.f.** T is the time for the e.m.f. to complete one cycle, that is to increase from zero to its maximum positive value, decrease from that point to its maximum negative value, and then return to zero. The number of times this occurs in a second is known as the **frequency** of the alternating e.m.f. and is given by

$$f = \frac{1}{T}$$

Figs. 17:2(a) and (b) show a sinusoidal variation of current and e.m.f. with time. Each can be written mathematically as follows:

$$E = E_0 \sin \omega t \qquad\qquad (17:1)$$

and $\quad I = I_0 \sin \omega t$

where E is the e.m.f. at a time t, and E_0 is the peak e.m.f.

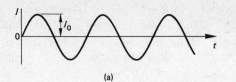

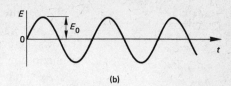

(a)   (b)

Fig. 17:2 (a) Sinusoidally varying current of the form $I = I_0 \sin \omega t$. (b) Sinusoidally varying e.m.f. of the form $E = E_0 \sin \omega t$.

I is the current at a time t, I_0 is the peak current, and ω is the angular velocity or angular frequency which is given by

$$\omega = 2\pi f$$

where f is the frequency of the alternating e.m.f. or current.

It should be noted that if alternating potential differences are being considered then the expression for the p.d. at any instant is usually given by

$$V = V_0 \sin \omega t$$

Example 17:1

The frequency of a sinusoidally varying e.m.f. is 50 Hz. If the peak e.m.f. is 300 V and at a time $t = 0$ the e.m.f. is zero, calculate the e.m.f.

    a) after 0.005 sec   *b*) after 0.01 sec   *c*) after 0.013 sec

Solution   Using eqn. (17:1),

$$\omega = 2\pi f = 2 \times \pi \times 50 = 314 \text{ rad/sec}$$

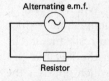

Alternating e.m.f.

Resistor

Fig. 17:3 Sinusoidally varying e.m.f. applied to a resistor.

a) $E = 300 \times \sin(314 \times 0.005) = 300 \times 1 = 300 \text{ V}$
b) $E = 300 \times \sin(314 \times 0.01) = 300 \times 0 = 0 \text{ V}$
c) $E = 300 \times \sin(314 \times 0.013) = 300 \times -0.809 = -243 \text{ V}$.

17:2 Sinusoidally-varying E.M.F. Applied to a Resistor, Capacitor and Inductor

1 Fig. 17:3 shows a sinusoidally varying e.m.f. applied to a **resistor**. Assuming that the resistor obeys Ohm's Law, then the current flowing through the resistor is proportional to the potential difference across the resistor. When a sinusoidally varying source of e.m.f. is applied across the resistor, the current through the resistor will be a maximum when the p.d. across it is a maximum (see fig. 17:4). The conclusion is that the current I must also be varying sinusoidally with time and is *in phase* with the p.d. across the resistor. (See section 6:6 for the idea of phase.) If the p.d. across the resistor is given by

$$V = V_0 \sin \omega t$$

then the current through the resistor is

$$I = I_0 \sin \omega t$$

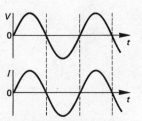

Fig. 17:4 P.d. across the resistor is in phase with the current through the resistor.

2 Fig. 17:5 shows a battery connected to a **capacitor**. The instant before the key is closed the potential difference across the plates of the capacitor is zero. Immediately the key is closed, charge starts to flow rapidly on to the plates and, as the p.d. across the plates increases, the rate of flow of

Capacitor

Battery   Key

Fig. 17:5 E.m.f. of a battery applied to a capacitor

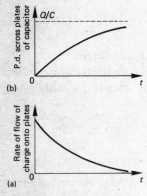

(b)

(a)

Fig. 17:6 (a) Increase of p.d. with time as a capacitor is charged; (b) Decrease of rate of change of charge flow with time as a capacitor is charged.

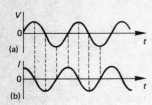

(a)

(b)

Fig. 17:7 (a) P.d. across capacitor changing with time and (b) the flow of charge changing with time. Note the phase difference: the current *leads* the p.d. by $\frac{1}{2}\pi$.

17:3 R.M.S. Potential Difference and Current

charge (the current) on to the plates decreases until eventually, when the p.d. across the plates is equal to the applied e.m.f., no further charge flows (see fig. 17:6(a) and (b)). When the p.d. across the plates is a maximum, the rate of flow of charge on to the plates is zero.

A similar variation of p.d. and current with time occurs when an alternating e.m.f. is applied to a capacitor. The p.d. varies as

$$V = V_0 \sin \omega t$$

and the rate of flow of charge on and off the plates of the capacitor is given by

$$I = I_0 \sin (\omega t + \tfrac{1}{2}\pi) \qquad (17:2)$$

Note that this is still a sinusoidal variation of current with time but that there is a phase difference of $\frac{1}{2}\pi$ between the current and p.d.. As the current reaches its maximum first (at $t = 0$), **the current leads the p.d.** by $\frac{1}{2}\pi$ (see fig. 17:7(a) and (b)).

Equation (17:2) can also be written as

$$I = I_0 \cos \omega t \qquad (17:3)$$

3 Fig. 17:8 shows a battery connected to an **inductor**. When the key is closed, the current increases from zero to its maximum value (which would be infinite if the inductor were ideal and had zero resistance). The greatest rate of change of current flow will occur when the key has just been closed. Therefore, at this instant, the back e.m.f. which opposes the increase in current will be at its maximum. As the current reaches its steady maximum value (which will be V/R where R is the resistance of the inductor windings), the back e.m.f. goes to zero. An important point to note is that, as the key is closed, the p.d. across the inductor is a maximum when the current through the inductor is zero as shown in fig. 17:9. This can be extended to an alternating e.m.f. applied to an inductor. Refer to fig. 17:10(a) and (b). As with the capacitor there is a phase difference between the p.d. across the inductor and the current through the inductor. If the p.d. across the inductor is given by $V = V_0 \sin \omega t$, then the current through the inductor can be shown to be

$$I = I_0 \sin (\omega t - \tfrac{1}{2}\pi) \qquad (17:4)$$

In this case the **current lags the p.d.** by $\frac{1}{2}\pi$. Equation (17:4) can be written as

$$I = -I_0 \cos \omega t$$

When a current flows through a resistor there is always a heating effect, whether the current is a.c. or d.c. When dealing with direct currents the power dissipated in the resistor is given by

$$P = I_{dc}^2 R$$

where I_{dc} is the d.c. current.

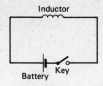

Fig. 17:8 E.m.f. of a battery applied to an inductor.

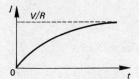

Fig. 17:9 Current versus time for an inductor just after the key is closed.

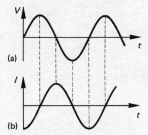

Fig. 17:10 (a) P.d. across inductor varying with time and (b) the current through the inductor changing with time. Note the phase difference: the current *lags* the p.d. by $\frac{1}{2}\pi$.

A similar expression can be derived for the power dissipated when an alternating current flows. As the value of the current changes over one cycle, so the power dissipated at any instant will also change. In order to be able to calculate the amount of energy that is converted from electrical energy to other forms of energy in some time interval it is necessary to know the *average* power dissipated.

$$\text{Average power } \langle P\rangle = \langle I^2\rangle R \qquad (17:5)$$

N.B. The brackets $\langle\ \rangle$ indicate the *average values* of these quantities over one cycle.

Now $\quad I = I_0 \sin \omega t$

so $\quad \langle P\rangle = \langle I_0^2 \sin^2 \omega t\rangle R = I_0^2\langle\sin^2 \omega t\rangle R$

(I_0^2 can be taken out of the brackets because it is constant.)

It can be shown mathematically that

$$\langle\sin^2 \omega t\rangle = \tfrac{1}{2}$$

Therefore,

$$P = \frac{I_0^2}{2} R = I_{rms}^2 R$$

where $\quad I_{rms} = \dfrac{I_0}{\sqrt{2}}$

I_{rms} is called the **root mean square value** of the alternating current. It is the value of a direct current which would produce the same heating effect in a circuit as an alternating current of peak value I_0. For example if the *peak* current through an a.c. circuit is 10 A, this would produce the same heating effect as $10/\sqrt{2}$ A of direct current flowing through the same circuit, i.e.

$$I_{rms} = \frac{10}{\sqrt{2}} \approx 7 \text{ A}$$

Similarly we can show that $\quad V_{rms} = \dfrac{V_0}{\sqrt{2}}$

The power dissipated in an a.c. circuit can be written as

$$P = V_{rms}I_{rms} = \frac{V_0}{\sqrt{2}} \times \frac{I_0}{\sqrt{2}} = \frac{V_0 I_0}{2} \qquad (17:6)$$

Example 17:2

Consider the circuit shown in fig. 17:11. If the peak current through the circuit is 5 A, calculate:

a) The r.m.s. current.
b) The peak p.d. across the resistor.
c) The r.m.s. p.d.
d) The power dissipated in the resistor.
e) The total energy converted in the resistor in 10 minutes.

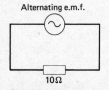

Alternating e.m.f.

10Ω

Fig. 17:11

Solution

a) $I_{rms} = \dfrac{I_0}{\sqrt{2}} = \dfrac{5}{\sqrt{2}}\,\text{A} = 3.54\,\text{A}$

b) $V_0 = I_0 R = 5 \times 10 = 50\,\text{V}$

c) $V_{rms} = \dfrac{V_0}{\sqrt{2}} = \dfrac{50}{\sqrt{2}}\,\text{V} = 35.4\,\text{V}$

d) $P = V_{rms} I_{rms} = \dfrac{50}{\sqrt{2}} \times \dfrac{5}{\sqrt{2}} = \dfrac{250}{2} = 125\,\text{W}$

e)   Total energy dissipated = average power × time
$$= 125 \times 10 \times 60 = 75\,000 \text{ joules} = 75\,\text{kJ}$$

Exercises

17.1 Sketch the phase difference between the p.d. across an inductor and the current through the inductor when an alternating e.m.f. is applied to the inductor.

17.2 If the peak current through an alternating current circuit is 12 A give the r.m.s. current.

17.3 The mains supply in Great Britain is 240 V r.m.s. Calculate
a) The peak p.d.
b) The peak-to-peak p.d.
If the mains supply is connected to a 2 kΩ resistor, calculate
a) The r.m.s. current through the resistor.
b) The peak current through the resistor.
c) The energy dissipated in the resistor in 10 minutes.

17.4 Explain how a source of alternating e.m.f. may be produced. State which factors affect the output e.m.f. of a dynamo.

18 The Electron and the Photon

The general ideas associated with atomic structure have already been indicated in several chapters of the text. The picture is of a massive positively charged nucleus, around which a cloud of negatively charged electrons exists, to provide an atom which is electrically neutral overall. In the following and final chapters of the text we shall explore these ideas in greater detail, study the ways in which models of the atom have arisen, and consider some applications and effects of this knowledge. We shall see that the emission and absorption of electromagnetic radiation are related to the outer electronic part of the atom and we shall also find that the nucleus is complex and often unstable. To begin with we shall consider the electron, its nature and behaviour as a free particle.

18:1 The Nature of Electrons

The concept of the free electron arose originally from the study of electric discharges in gases at low pressure. It was apparent that some form of radiation was emanating from the cathode of the gas discharge tube and, hence, electrons were initially called "cathode rays". The name still survives in the cathode ray oscilloscope (CRO). The early experimenters were not clear about the nature of the cathode rays and not until J. J. Thomson published his work were the rays identified properly as high-speed negative particles which we now call electrons.

Thomson originally investigated the nature of electrons using a *cathode-ray gas-discharge tube*. The experiment can be reproduced with a modern *oscilloscope tube* as shown in fig. 18:1. The coils either side of the tube, and the deflector plates within the tube, enable the beam of electrons to be subjected to electric and magnetic fields which are mutually perpendicular.

Fig. 18:1 The modern equivalent of the Thomson experiment for *e/m* of electrons.

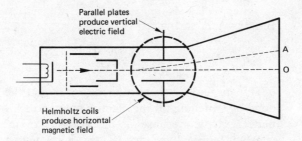

Parallel plates produce vertical electric field

Helmholtz coils produce horizontal magnetic field

At first the magnetic field only is applied and results in a deflection of the beam, as indicated by the spot on the screen moving from O to A. In passing through the magnetic field B, the electrons experience a force which bends them into a circular trajectory such that

$$Bev = \frac{mv^2}{R} \qquad (18:1)$$

where R = radius of the curved path, which can be calculated from the geometry of the tube, e is the charge on the electron, m is the mass of the electron, and v its velocity.

If now a potential difference is applied between the deflector plates, an electric field is set up which can be adjusted to return the beam to its original position at O. When this occurs, the electric and magnetic forces on the electrons must be just equal and opposite. In this situation we can write

$$Bev = eE \qquad (18:2)$$

From equation (18:1)

$$\frac{e}{m} = \frac{v}{BR}$$

and substituting for v from equation (18:2), i.e. $v = E/B$, we have

$$\frac{e}{m} = \frac{E}{B^2R}$$

e/m is known as the **specific charge** of the particle.

For the electron, the value for e/m is 1.76×10^{11} C/kg.

A direct measurement of the electron's charge was first carried out by Millikan in 1911. The basis of **Millikan's experiment** is indicated in fig. 18:2(a). Tiny drops of oil are sprayed from an atomizer into the top chamber and a few of these will drift down through the central hole into the space between the two parallel conducting plates. The oil drops carry charges generated by frictional effects at the atomizer nozzle. By applying a suitable potential difference—a negatively charged drop is illustrated, requiring the top plate to be made positive with respect to the bottom— one of the drops can be held stationary, with its weight mg exactly balanced by the upward force on the charge q produced by the electric field E (see Fig. 18:2(b)). In this situation

$$qE = mg$$

But $E = V/d$, so that

$$q\frac{V}{d} = mg \qquad \text{i.e.} \quad q = \frac{mgd}{V}$$

Thus to find q it is also necessary to find the mass of the drop. This is done by switching off the electric field and allowing the drop to fall freely under gravity. Because of the viscosity of the air, the drop reaches a constant terminal velocity and this is related to the mass of the drop, the viscosity of the air, and the density of the oil in the drop.

By repeatedly raising the drop, using the electric field, and letting it fall

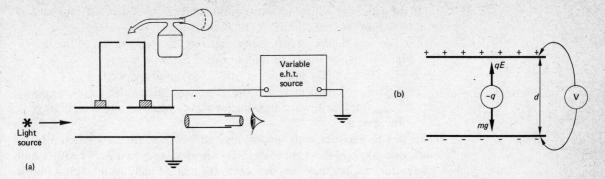

Fig. 18:2 (a) Schematic diagram of the Millikan experiment for measuring the electronic charge. (b) Forces acting on a charged drop between the plates.

under gravity, a good estimate of the terminal velocity, and hence the mass of the drop, can be made. Since d and V can be measured easily and g is a known constant, the charge carried by any given drop can be determined.

On repeating the experiment with many different drops, it is found that q always has the value 1.6×10^{-19} C or some integral multiple of this value. Consequently this must correspond to the fundamental unit of electrical charge, i.e. the charge on the electron e is 1.6×10^{-19} C.

18:2 Planck's Quantum Theory

We have already seen in chapter 7 how phenomena such as interference, diffraction and polarization suggested that light must have the nature of a wave motion. We have also seen that light is only a small part of a large spectrum of electromagnetic radiation. One aspect of the electromagnetic spectrum which was of particular interest to physicists around the turn of the century was **black-body radiation**.

A close approximation to a black-body in the laboratory is a small hole in the side of a hollow chamber, which has a matt black surface on its inner walls, as illustrated in fig. 18:3. If the whole chamber is now heated, the small hole will emit radiation which is very nearly black-body radiation, i.e. it contains nearly all possible wavelengths of the electromagnetic spectrum.

Fig. 18:3 Laboratory simulation of a black body

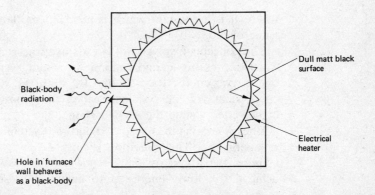

If the chamber is heated to different temperatures and the intensity of the radiation is measured, it is discovered that:

1   The total energy emitted per unit time is proportional to the fourth power of the absolute temperature: *Stefan's law*.
2   The black-body radiation forms a continuous spectrum.
3   The wavelength at which the radiation carries the greatest energy (λ_{max}) moves toward the shorter wavelengths as the temperature is increased.

This last fact agrees with simple observation. For instance, when a body such as a piece of steel is progressively heated, at first it glows with a dull cherry red colour, then orange, then white, and finally white with a bluish tinge. These facts are illustrated in fig. 8:8 (p. 85).

Physicists such as Rayleigh, Jeans and Wien attempted to develop theories which would explain the distribution of energy in the black-body radiation spectrum. Their theories were all based on ideas involving tiny atomic oscillators which continuously absorbed energy from the source of heat and continuously re-emitted it as electromagnetic radiation over the whole spectrum. Unfortunately this approach did not produce a proper explanation.

However in 1900, the whole edifice of classical physics received a shattering blow. It came in the form of a suggestion by Max Planck who proposed a startlingly new model for the emission and absorption of energy which was able to predict theoretically the precise form of the black-body radiation curve.

Planck abandoned the idea of continuous emission and absorption of energy by the atomic oscillators and proposed instead that the process was *discontinuous*. The energy was emitted in discrete amounts which Planck called quanta. He further proposed that the energy associated with any **quantum** of radiation was proportional to its frequency. Mathematically this may be expressed as

$$E \propto f$$

and then by introducing a constant this becomes an equality:

$$E = hf$$

where h is a constant which is now called **Planck's constant**. It has the value of 6.63×10^{-34} J s.

The concepts involved in Planck's hypothesis are illustrated in fig. 18:4 which represents a simple atomic oscillator system able to exist in only one of two energy states A and B, where $E_B > E_A$. The lowest possible energy state of the system is called the **ground state** and the allowed higher-energy state the **excited state**.

Energy coming in is absorbed and the system moves to its excited state of energy E_B. This condition is unstable and, as in all known physical systems, the atomic oscillator tends to its lowest-energy state. In order to achieve the lower-energy state again, the oscillator must release an

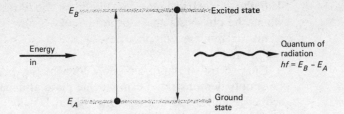

Fig. 18:4 Illustration of a simple atomic system in terms of the Planck hypothesis.

amount of energy $E_B - E_A$ and this appears as a quantum of radiation of frequency f given by Planck's law:

$$E_B - E_A = hf \qquad\qquad (18:3)$$

This new thinking not only solved the black-body radiation problem but formed one of the fundamental building blocks of modern physics.

Example 18:1

Determine the energy of a quantum of radiation of wavelength 590 nm.

Solution   For a quantum of radiation of frequency f and wavelength λ we have

$$E = hf = \frac{hc}{\lambda}$$

where $c = 3 \times 10^8$ m/s, the velocity of electromagnetic radiation in a vacuum. Therefore

$$E = \frac{6.63 \times 10^{-34} \times 3 \times 10^8}{590 \times 10^{-9}} = 3.37 \times 10^{-19} \text{ J}$$

This is a very small amount of energy and would be capable only of lifting a small fly through a vertical distance of one millionth of a micron.

As the wavelength decreases, the energy of the quantum increases rapidly. The relative energy content, effects and origins are illustrated in fig. 18:5.

Fig. 18:5 Relative energies of quanta across the electromagnetic spectrum.

SPECTRUM BAND	Infra-red	Visible	Ultra-violet	X-ray	Gamma-ray
Typical wavelength	1234 nm	620 nm	205 nm	0.124 nm	0.00124 nm
Approximate energy	1 eV	2 eV	6 eV	10000 eV	1000000 eV
Illustration of relative energies in terms of missiles	Air gun pellet	.22 rifle	.303 rifle	4.5 inch shell	15 inch shell

18:3 The Photo-electric Effect

The photo-electric effect was, like black-body radiation, another problem which the physicists around the turn of century were unable to explain in terms of the existing theories of the wave nature of electromagnetic radiation.

The process of photo-emission, for which explanation was sought, may be simply demonstrated using a zinc plate attached to a gold-leaf electroscope as illustrated in Fig. 18:6. When UV light is incident on the zinc plate, the collapsing leaf indicates that electrons are being lost from the zinc plate.

The phenomenon itself may be studied in more detail using apparatus similar to that shown schematically in fig. 18:7. Using such apparatus various basic facts can be established:

1 Photo-electrons are emitted with various kinetic energies up to a maximum value, and this maximum value is dependent on the incident light frequency (not the intensity, as classical electromagnetic theory implies).

2 The number of photo-electrons emitted is proportional to the intensity of the light.

3 No photo-electrons are emitted below a certain threshold frequency which is different for different conductors.

4 Photo-electrons appear instantaneously with the illumination of the surface.

It was in 1905 that Einstein realized the significance of Planck's ideas with respect to the photo-electric effect. He extended the concept whereby the atoms simply emitted their energy in discrete quanta. He proposed that, having been emitted, this quantum remained as a unit packet of energy travelling through space at the velocity of light. These travelling packets of electromagnetic energy then came to be called **photons**. The energy of a photon is given as before by Planck's law: $E = hf$.

What Einstein proposed was that the photon penetrated a few atomic layers into the material and interacted with one of the "free" electrons in the conductor. In this single sudden event, the whole of the photon energy is transferred to the electron as kinetic energy. If the electron gains sufficient energy and has its velocity in the right direction, it will appear outside the conductor with a maximum kinetic energy given by the photon energy *less* the amount of work done to overcome the electrostatic retaining forces (see fig. 18:8).

This process is simply represented by the equation:

$$(\tfrac{1}{2}mv^2)_{max} = hf - \phi \qquad\qquad (18:4)$$

where ϕ is the work function, which is a measure of the amount of energy that the electron must expend in order to escape from the electrostatic forces that normally keep it within the metal. It has a value which is different for different metals and depends also on the physical state of the surface.

This equation is known as Einstein's **photo-electric equation** and all the observed features of the photo-electric effect are fully described by it.

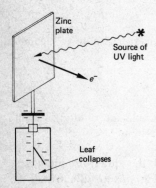

Fig. 18:6 Demonstration of the photo-electric effect.

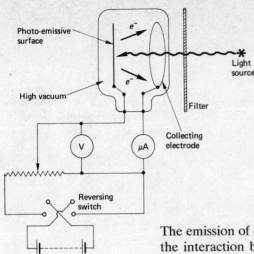

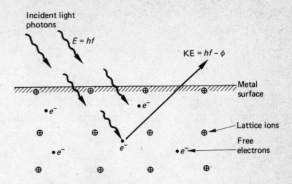

Fig. 18:8 Illustration of the photo-electric effect

Fig. 18:7 Apparatus to observe the process of photo-electric emission.

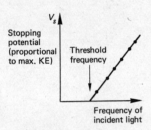

Fig. 18:9 Graph of stopping-potential against frequency of incident light.

The emission of electrons at lower than the maximum energy occurs when the interaction between the photon and the electron takes place deeper inside the metal, and therefore the electron must do more work in order to escape. The maximum energy of the emitted electrons can be measured in terms of the *stopping potential* V_s. This is the reverse potential applied to the collecting electrode just necessary to prevent the collection of electrons indicated by zero current on the microammeter. In this situation the initial kinetic energy of the emitted photo-electron is just used up in doing work against the adverse electric field. Thus we may write

$$(\tfrac{1}{2}mv^2)_{max} = eV_s \qquad \text{(using equation 13:8)}$$

and the Einstein equation becomes

$$eV_s = hf - \phi$$

Dividing through by e then gives

$$V_s = \frac{h}{e}f - \frac{\phi}{e}$$

This equation is now in linear form and a plot of stopping potential V_s against incident light frequency should give a straight line of slope h/e. This is indeed the case and, in fact, was initially used to obtain an experimental determination of Planck's constant h (see fig. 18:9).

The threshold frequency is also simply explained since no emission can take place unless

$$hf \geqslant \phi \qquad \text{i.e. unless } f \geqslant \frac{\phi}{h}$$

Different materials with different work functions will then give rise to different threshold frequencies, which is another observed fact.

Finally, since each interaction between a photon and electron is a single sudden event, the number of released electrons must be proportional to the number of incident photons, i.e. the intensity of the light. Also, *instantaneous* emission would be expected, since no extended period of time is needed for the electron to absorb sufficient energy to escape (as was predicted by the classical wave theory).

Example 18:2

When light of wavelength 650 nm is incident on a metal surface photo-electrons are emitted with a maximum velocity of 3.6×10^5 m/s. Determine *a*) the work function in electron-volts and *b*) the threshold frequency.

Solution

a) $KE = \frac{1}{2}mv^2$

$$= \frac{1}{2} \times 9.1 \times 10^{-31} \times (3.6 \times 10^5)^2 \, J = 0.59 \times 10^{-19} \, J$$

Hence, using the Einstein equation:

$$0.59 \times 10^{-19} = \left(6.6 \times 10^{-34} \times \frac{3 \times 10^8}{650 \times 10^{-9}}\right) - \phi$$

$$\therefore \qquad \phi = 2.46 \times 10^{-19} \, J$$

But $1 \, eV = 1.6 \times 10^{-19} \, J$ and hence this becomes $\phi = 1.54 \, eV$

b) Threshold frequency $f = \dfrac{\phi}{h}$

$$= \frac{2.46 \times 10^{-19}}{6.6 \times 10^{-34}} = 3.72 \times 10^{14} \, Hz$$

18:4 Wave/Particle Duality

The photo-electric effect can only be accounted for if we consider light to have particle-like properties. It is necessary to visualise a light beam being made up from a stream of photons all travelling at the velocity of light. At the same time we are aware that light is a wave motion because it demonstrates wave-like behaviour in such phenomena as interference, diffraction and polarization. Hence we see a **dual nature**—light appears both as a **particle** and a **wave**—something which seems very odd.

This fact only seems strange, however, because of our preconceived ideas of how waves and particles should behave. We think entirely in terms of such things as billiard balls and waves on strings, phenomena which are common in our everyday experience. We see these as entirely separate and independent effects.

However, particles and waves are, in fact, very much alike since they represent the two possible means of transferring energy between two points in space, i.e. they are simply two different forms of the same process, namely energy transport.

Thus the wave and particle concepts are **complementary** and both are required to describe the phenomena. Effectively we use

> the particle model for emission and absorption of electromagnetic energy
> the wave model for propagation through space.

Exercises

Data $e = 1.6 \times 10^{-19}$ C $\qquad c = 3 \times 10^8$ m/s $\qquad m_e = 9.1 \times 10^{-31}$ kg

$g = 9.8$ m/s^2 $\qquad h = 6.63 \times 10^{-34}$ J s

18.1 In Millikan's experiment an oil drop of radius 5 μm and density 0.8×10^3 kg/m^3 is in equilibrium between the plates when it carries a charge equivalent to two excess electrons. Determine the potential difference between the plates if they are 5 mm apart.

18.2 Calculate the mass of an oil drop carrying five electrons which is just balanced between two plates 1.5 cm apart if there is a potential difference of 2000 V across them.

18.3 Calculate the energy in electron-volts of a photon of wavelength 10 nm.

18.4 The energy of a gamma-ray is 5 MeV. Calculate the wavelength.

18.5 Calculate the voltage required to stop a photo-electron ejected from a metal surface at a velocity of 10^6 m/s.

18.6 The work function of a metal is 2.2 eV. Calculate the maximum kinetic energy of electrons ejected from this surface by radiation of 240 nm.

18.7 Determine the threshold frequency for a tungsten surface which has a work function of 4.53 eV.

18.8 Red light of wavelength 675 nm is just able to liberate electrons from a metallic sodium surface. Calculate the work function of sodium in electron-volts.

18.9 Decide whether the yellow light from a sodium vapour lamp ($\lambda = 589.3$ nm) can eject photo-electrons from a clean metallic surface with a work function of 2.46 eV.

18.10 Determine the wavelength of radiation that will eject electrons from a metallic surface with a maximum kinetic energy of 2.5×10^{-19} J if the work function of the surface is 2.3 eV.

19 A Model for the Atom and Emission Spectra

Our knowledge of the behaviour of the outer atom has come from an analysis of the light, i.e. photons, emitted and absorbed by the atom. The outer part of the atom consists of a number of electrons which are in constant motion about the central nucleus. Study of the light leads us to the concept of energy levels within the atom and an orbital model where the atom is seen as a miniature solar system—electrons as the planets and the nucleus as the sun. This model is now known to be only a special case of a more general theory but nevertheless provides us with a valuable insight into atomic behaviour. We will again approach the subject historically, starting with Rutherford's experiments in which high-speed alpha-particles are scattered by atoms of a gold foil.

19:1 The Charge Distribution within the Atom

At the turn of the century it was clearly understood that electrons must form part of the atom and that they were also related in some way to the light emitted by the atom. Also, the fact that the atom as a whole was electrically neutral required there to be present in the atom some positive charge. The view of the atom held at that time was proposed by J. J. Thomson where the electrons were embedded in some sort of positive material, which was not clearly defined (see fig. 19:1). This led to the nickname "currant-bun atom", the most significant aspect of which was that the positive charge was distributed over the whole volume of the atom, which had a radius inferred from chemical analysis of crystal structures to be of the order of 10^{-10} m.

Rutherford proposed a new model of the atom following an analysis of experiments which involved high-speed alpha-particles from a radioactive isotope being directed at a thin gold foil. Alpha-particles (or α-particles) are nuclei of helium (charge $+2e$) which are ejected at high speed from certain unstable atoms. Their properties will be considered in chapter 21. The experiments were carried out by Geiger and Marsden in 1909 and the form of their apparatus is shown in fig. 19:2.

With the currant-bun atom, all the alpha-particles would suffer *some* deflection when they passed near to or through the areas of positive charge according to Coulomb's law of force between electrostatic charges (see chapter 13). However at no time are the alpha-particles close to any concentrated positive charge and therefore they would not be expected to suffer large deflections, as illustrated in fig. 19.3(a). The experimentation of Geiger and Marsden gave a totally different result. They found that most of the alpha-particles went straight through the foil undeviated but a few suffered large deflections which in some cases were greater than 90° (see fig. 19:3(b)).

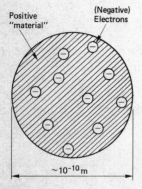

Positive "material"

(Negative) Electrons

$\sim 10^{-10}$ m

Fig. 19:1 The currant-bun atom.

Fig. 19:2 The alpha scattering experiment of Geiger and Marsden.

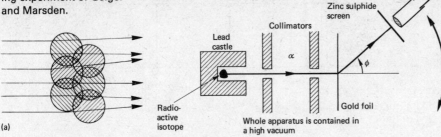

(a)

(b)

Fig. 19:3 Deflection of alpha particles by (a) currant-bun and (b) Rutherford atom.

To explain the necessarily large repulsive forces which would be required in order to deflect the alpha-particles in such a dramatic way, Rutherford was forced to envisage a tiny region of concentrated positive charge, which he called the **nucleus**, existing within the atom. Using only the Coulomb law of force between the positive nucleus and the positive alpha-particle, Rutherford was able to show that the results of the experiments were consistent with a single scattering event from an extremely small nucleus. He was able to estimate an upper limit to its size—something of the order of 10^{-14} m. Hence Rutherford was led to the **nuclear atom**: a tiny nucleus containing all the positive charge of the atom and most of its mass, surrounded by largely empty space except for the electrons needed to keep the whole electrically neutral.

To appreciate the relative magnitudes involved here, we can imagine the whole atom to be spread over the area of a football pitch. The nucleus would then be seen as a golf-ball on the centre spot. The alpha-particle which could be represented by a pea would then have to be projected across the pitch and, in order to suffer any deflection, would need to pass within 50 cm of the golf-ball. It is therefore not very surprising that only a few alpha-particles suffered large deflections.

19:2 Emission Spectra

Concurrent with Rutherford's work on the nuclear atom and for some time before, a considerable amount of research into atomic spectra had been carried on by various workers. The fundamental problem which these spectroscopists were seeking to explain was the significance of the unique line emissions associated with each of the various elements of the periodic table. These spectra and their production have already been discussed in chapter 8 (sections 8:1, 8:2, 8:3).

Because of the great complexity of most emission spectra, work tended to concentrate on the simpler spectra and eventually the simplest of all, namely that of hydrogen—the simplest element of all. It was considered that, if the **spectrum of hydrogen** could be related to its simple structure, understanding of the more complex spectra would follow. The visible emission spectrum of hydrogen is illustrated in fig. 19:4. It is obvious that there is an underlying simplicity. The lines tend to get progressively closer

Fig. 19:4 The visible spectrum of hydrogen.

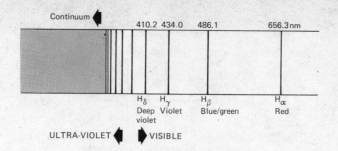

together from the red H_α until they merge entirely in a region which is referred to as the *continuum* and lies in the near ultra-violet.

The breakthrough occurred in 1885 when Balmer produced a formula which could be used to predict all the lines of the visible hydrogen spectrum to within ±4 Å. The formula Balmer obtained may be written as

$$\frac{1}{\lambda} = R\left(\frac{1}{2^2} - \frac{1}{n^2}\right)$$

where n is an integer and R is a constant called the Rydberg constant, after another spectroscopist who demonstrated that similar relations existed for other spectra.

Subsequently other series of emission lines from hydrogen were discovered, lying in the ultra-violet and infra-red regions of the electromagnetic spectrum. For example

Lyman series   UV
Balmer series   Visible/near UV
Paschen series   IR
Bracket series   IR

However the underlying problem is still present in that these formulae are all empirical and have no theoretical basis. Why should excited atoms emit specific lines unique to each element, and why in the case of hydrogen can they be described by formulae of the type discovered by Balmer?

19:3 The Bohr Model of the Hydrogen Atom

The Rutherford model of the atom with its tiny massive positive nucleus and associated negative electron immediately raises further problems. What happens to the electron? It cannot remain stationary as it would "fall" into the nucleus owing to the mutual electrostatic attraction. Alternatively, if it orbits around a circular path, the necessary centripetal acceleration would cause it to radiate electromagnetic energy. (Whenever a charge is accelerated, radiation is emitted, a phenomenon used by all radio transmitters.) The effect of this loss of energy would again cause the electron to spiral into the nucleus as its energy progressively decreased. Obviously none of these things happen. We know that atoms are basically stable and do not progressively lose energy, emitting radiation continuously at wavelengths right across the electromagnetic spectrum.

A new understanding of the form and behaviour of the atom began following a series of postulates made by a physicist called Bohr in 1913. Bohr's suggestions permitted the observed spectrum of hydrogen to be related directly to a particular model of the atom and also provided a theoretical basis for the relations discovered by Balmer sometime earlier. The theory was so successful in predicting the spectral wavelengths that physicists realized it heralded a new age of understanding concerning the mysteries of the atom. The theory is a mixture of well-established classical mechanics and the new quantum theory introduced by Planck, and it starts from a basis of the Rutherford atom.

The model of the hydrogen atom proposed by Bohr is described entirely within the postulates which follow:

1 The atom can exist in any one of a number of radiationless states, $E_1, E_2, E_3, \ldots, E_n$, which correspond to the electron being in one of a number of stable circular orbits around the nucleus.

In this postulate, Bohr is using the form of the Rutherford atom with rotating electrons and assumes that energy is neither emitted nor absorbed, contrary to the accepted principles.

2 When an electron makes a transition from one orbit of energy E_2 to another of lower energy E_1, then a quantum of electromagnetic energy is emitted given by

$$E_2 - E_1 = hf$$

This postulate uses the already established theory of Planck (see chapter 18) regarding the emission and absorption of radiation.

3 The allowed orbits are defined by the quantum condition that the angular momentum of the electron must be an integral multiple of Planck's constant h, divided by 2π, i.e.

$$\text{Angular momentum} = n \times \frac{h}{2\pi}$$

where $n = 1, 2, 3, \ldots$.

This third postulate was the particularly original contribution of Bohr. What this does is to quantize the angular momentum into certain allowed values only, which in turn define the radius of the orbit and hence also the energy of the particular allowed state.

4 The stable orbits are determined by equating the centripetal acceleration force to the Coulomb attraction between the nucleus and the electron.

This final postulate is simply a statement of well-established electrostatic and mechanical principles.

What is expressed in words in these postulates may very simply be

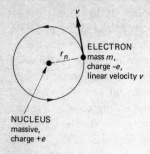

ELECTRON
mass m,
charge $-e$,
linear velocity v

NUCLEUS
massive,
charge $+e$

Fig. 19:5 The electron in orbit.

written down in the form of two equations. From the fourth postulate we may write

$$\frac{e^2}{4\pi\epsilon_0 r_n^2} = \frac{mv^2}{r_n} \qquad (19:1)$$

where r_n represents the radius of the stable orbit corresponding to the number n.

Considering the third postulate, we see from fig. 19:5 that the angular momentum is given by

$$L = I \times \omega \qquad \text{(see section 4:5)}$$

For a rotating point mass, the moment of inertia is mr_n^2, hence

$$L = (mr_n^2) \times \left(\frac{v}{r_n}\right) = mvr_n$$

Hence a second equation can be written down:

$$mvr_n = \frac{nh}{2\pi} \qquad (19:2)$$

When these two equations are solved, a very clear picture of the hydrogen atom unfolds. To begin with it is discovered that the radii of the allowed stable orbits are found to be proportional to the square of n, i.e.

$$r_n \propto n^2$$

This means that, as the orbit number n changes, the distance of the electron from the nucleus rapidly becomes very large.

Also, when the total energy of the electron is considered, made up partly as kinetic energy due to its orbital velocity and partly as electrostatic potential energy in the field of the positive nucleus, the energy is found to be inversely proportional to the square of n, i.e.

$$E_n \propto \frac{1}{n^2}$$

This total energy E_n is also found to be negative, indicating a "bound" state. This means that the electron is tied to its particular nucleus by the electrostatic field. The electron is said to be in a negative potential well. The most negative state, i.e. the most tightly bound electron, occurs for $n = 1$. This is referred to as the **ground state** of the atom.

The value of n is seen to be of great significance in controlling the state and energy of the atom and for this reason is given the name **principal quantum number**. As already seen, it can only have the integral values $1, 2, 3, 4, \ldots$ up to infinity. Thus the form of the hydrogen atom implied by the Bohr theory is as illustrated in fig. 19:6. The diagrams show the relative positions of the orbits around the nucleus and their corresponding energies.

By introducing a constant A, the expression for the **total energy** may be written as

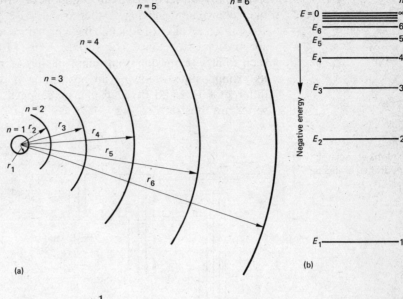

Fig. 19:6 (a) The form, and (b) energy levels, of the Bohr atom.

$r_2 = 4r_1$
$r_3 = 9r_1$
$r_4 = 16r_1$
$r_5 = 25r_1$
$r_6 = 36r_1$

(a)

(b)

$$E_n = -A \times \frac{1}{n^2}$$

Suppose the electron were to change orbits from one defined by principal quantum number n_2 to one of principal quantum number n_1, where n_2 is greater than n_1. The corresponding energy of the two states involved are then given by

$$E_1 = -A \times \frac{1}{n_1^2} \quad \text{and} \quad E_2 = -A \times \frac{1}{n_2^2}$$

Hence the change in energy which occurs as a result of the transition will be ΔE where

$$\Delta E = E_2 - E_1 = -A\left(\frac{1}{n_2^2} - \frac{1}{n_1^2}\right) = A\left(\frac{1}{n_1^2} - \frac{1}{n_2^2}\right)$$

But from the Planck law we may write $\Delta E = hf$ so that

$$hf = A\left(\frac{1}{n_1^2} - \frac{1}{n_2^2}\right)$$

$$h\frac{c}{\lambda} = A\left(\frac{1}{n_1^2} - \frac{1}{n_2^2}\right)$$

$$\frac{1}{\lambda} = B\left(\frac{1}{n_1^2} - \frac{1}{n_2^2}\right)$$

where B is another constant. This final equation represents the success of the theory since we see that it is identical to the original Balmer formula and the constant B corresponds to the Rydberg constant. The value of the constant obtained from the theory is found to agree very well with that determined from spectroscopic data.

209

Thus the model of the hydrogen atom described by Bohr is a considerable achievement, giving a theoretical base to the arbitrary Rydberg constant and also producing the correct relationship for the observed spectral emission lines. The various series of emission lines can now be seen to arise from quantum jumps of the electron as the atom returns to its ground state following an absorption of energy. The origin of the different series and detailed energy changes in the case of the Balmer series are illustrated in fig. 19:7.

Fig. 19:7 Energy levels and transitions associated with the hydrogen emission series.

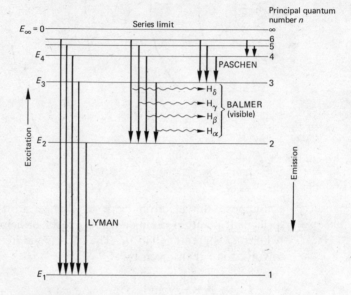

Example 19:1

The fourth line in the Balmer series of hydrogen (H_δ) has a wavelength of 410 nm. Identify the electronic transition which produces it and hence obtain an estimate of the Rydberg constant.

Solution Referring to fig. 19:7 we see that the H_δ line corresponds to a jump from orbit $n = 6$ to $n = 2$. Hence using the relation:

$$\frac{1}{\lambda} = R\left(\frac{1}{n_1^2} - \frac{1}{n_2^2}\right)$$

we have on substituting these values and also the wavelength of the H_δ line

$$\frac{1}{410 \times 10^{-9}} = R\left(\frac{1}{4} - \frac{1}{36}\right) = R\left(\frac{8}{36}\right)$$

$$R = \frac{36 \times 10^9}{8 \times 410} = 1.097 \times 10^7 \text{ per metre}$$

19:4 Excitation and Ionization

Using the concepts of the Bohr model we can describe the processes involved when light is emitted from a gas discharge tube. Free electrons in the gas are accelerated in the electric field created between the electrodes of the tube (see fig. 8:3). As the electrons gain velocity, they suffer **collisions** with the gas atoms in the tube. Initially these collisions are perfectly elastic and therefore just result in an exchange of kinetic energy. However, as the electron velocity increases, a point is reached when the electron kinetic energy is just equal to that required to raise the gas atom to its first **excited state**, e.g. $n = 2$ in the case of hydrogen. At this point the collision becomes inelastic and the electron kinetic energy is absorbed by the atom—a process referred to as **excitation**.

If the electron strikes the gas atom with kinetic energy greater than that required to excite the first level, the atom may be raised to even higher levels corresponding to $n > 2$. However, the energy absorbed can only match *precisely* the possible quantum difference between the states involved. When the energy supplied lies between two possible excited states, the atom absorbs only that which is necessary to bring about the permitted quantum change, while the rest is carried away with the electron.

In the situation where the incoming energy exceeds that necessary to raise the atom to the excited state corresponding to $n = \infty$, the atomic electron moves to an orbit of radius infinity (effectively this means something greater than about 10^{-8} m). Its total energy then becomes zero and the electron is no longer bound to its particular nucleus and is therefore free. With one electron missing from an atom, the atom is no longer electrically neutral and is referred to as a **positive ion**. This process is called **ionization**.

In the case of hydrogen (see fig. 19:8), the energy required for ionization is seen to be 13.6 eV and is just equal to the photon energy of the most energetic line of the Lyman series. Suppose a hydrogen atom received 13.9 eV following a collision. The atom is ionized and the free electron flies away with 0.3 eV of kinetic energy. We call 13.6 V the *ionization potential* of hydrogen.

Fig. 19:8 Energy levels of the hydrogen atom.

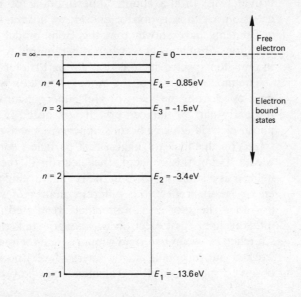

$n = \infty$ --- $E = 0$ --- Free electron

$n = 4$ $E_4 = -0.85$ eV

$n = 3$ $E_3 = -1.5$ eV   Electron bound states

$n = 2$ $E_2 = -3.4$ eV

$n = 1$ $E_1 = -13.6$ eV

Fig. 19:9 Three possible modes by which the atom might return to its ground state, producing a different set of line emissions in each case.

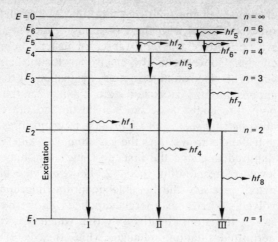

Once the atom has been excited or ionized it is in a very unstable condition. It tries to return to its ground state as quickly as possible. This typically takes place in times of the order of 10^{-8} sec. The electron can return from an excited state by a single quantum jump or by any possible sequence of jumps. Three possible modes are illustrated in fig. 19:9. Each group of transitions gives rise to a different set of emission lines. With the extremely large number of gas atoms involved, each of which is being ionized or excited to different levels, it is not surprising that when we examine the light from the tube a full spectrum is seen, containing all possible lines corresponding to all possible quantum jumps.

19:5 The Bohr Model and Quantum Theory

The Rutherford–Bohr model of the atom has produced a remarkably good explanation of the observed emission spectrum of hydrogen. It provides us with a sound understanding of the general processes of emission and absorption of energy by atoms. However, subsequent developments have shown that the Bohr model cannot be applied to all situations. Problems arise when it is required to account for fine structure in emission spectra such as doublet and triplet lines.

The modern theory of the atom is based on **wave mechanics** which was developed from the ideas of Heisenberg, Schrödinger and de Broglie in the early part of this century. It was noted in section 18:4 that, in the photo-electric effect, photons appear to behave as particles. De Broglie seized on this idea to suggest that particles, conversely, might behave as waves. Using this concept, the electron in the atom must now be visualized as a standing wave in the region around the nucleus. The different energy levels reflect the different modes of standing wave which are possible. The allowed vibrations are restricted to those modes having an integral number of electron wavelengths (c.f. standing waves on a string stretched between two rigid supports, see section 6:9). When the theory is worked out in detail, all the energy levels necessary to account for the observed spectra of all the elements can be predicted.

With the electron no longer viewed in terms of a small point mass moving in orbit around the nucleus, but rather by a three-dimensional vibration in space, it is not possible to state precisely where the electron is at any moment in time. So, when using the wave mechanics model, it is necessary to work with *mathematical probabilities* rather than precisely located circular orbits.

In conclusion it is simply necessary to note that obviously the Bohr model cannot be seen as wholly accurate in describing the form and behaviour of atoms. This must be left to the more sophisticated wave mechanical theory. Nevertheless, it is a most valuable model giving a good clear insight into the quantization of energy levels within the atom and the relation of these levels to the observed emission spectra. Provided its limitations are recognised and understood, the Bohr model offers a very useful and visual concept of atomic behaviour.

Exercises

Data $e = 1.6 \times 10^{-19}\,\text{C}$ $c = 3 \times 10^{8}\,\text{m/s}$ $\epsilon_0 = 8.85 \times 10^{-12}\,\text{F/m}$
$h = 6.63 \times 10^{-34}\,\text{J s}$ $R = 1.097 \times 10^{7}\,\text{per m}$

19.1 Write down the two equations which represent the Bohr postulates for hydrogen. Solve the equations for the radii of the stable orbits (r_n) and hence obtain a value for the radius of the first Bohr orbit.

19.2 Using the value of the radius calculated in question 1, determine the electric potential in volts at a point in the first Bohr orbit of the hydrogen atom.

19.3 Find the wavelength of the H_γ emission in the Balmer series of hydrogen.

19.4 Use the energy level diagram of fig. 19:8 to compute the longest wavelength line in the Lyman series of hydrogen. Is this line in the visible region of the electromagnetic spectrum?

19.5 Calculate, using the Rydberg formula, the wavelengths of the first three lines in *a*) the Balmer series and *b*) the Paschen series.

19.6 Determine the energy required to ionize an electron from the excited state corresponding to $n = 4$ in the hydrogen atom.

20 The Nucleus

In chapter 19 we saw the development of atomic theory from the original ideas of Rutherford. Subsequently Bohr developed his model of the hydrogen atom which incorporated the Rutherford ideas and envisaged electrons in certain stable orbits about the nucleus corresponding to quantized energy levels.

The development of the theory of the nucleus has been largely independent of atomic theory. The forces and energies involved are found to be immensely greater than those for atomic phenomena. However, to understand nuclear behaviour, it is necessary again to involve the classical laws of the conservation of energy and momentum and also the ideas of quantum theory. Experimental and theoretical research is still being carried out on the nucleus and there is still much to be properly understood. In this chapter a review of the composition, properties and behaviour of the nucleus which are generally understood and accepted will be given.

20:1 Composition and Size of the Nucleus

The nucleus itself has a complex structure but two types of nuclear particle called **nucleons** are known to exist within it. These nucleons are identified as *protons* and *neutrons* and there is abundant experimental evidence to suggest that combinations of these two nuclear particles give rise to all the different nuclear species known, both stable and unstable.

The masses of the proton and neutron are similar, with the neutron being slightly the heavier:

$$m_p = 1.672\,614 \times 10^{-27}\,\text{kg}$$

$$m_n = 1.674\,920 \times 10^{-27}\,\text{kg} \quad \text{(which is approximately 1800 times heavier than the electron).}$$

The **proton** carries one unit of positive charge (equal to that on an electron but of opposite sign, i.e. $+1.6 \times 10^{-19}$ C). The **neutron**, however, is electrically neutral and carries no charge at all. Since the atom is electrically neutral overall, the number of protons contained in the nucleus exactly equals the number of electrons in the extra-nuclear structure. However the total mass of the atom is governed by the number of neutrons associated with these protons. A typical element such as lithium has an atom which consists of a nucleus containing 3 protons and 4 neutrons, with 3 electrons in orbit around it as shown in fig. 20:1.

The number of protons present in any nucleus is called the **atomic number** (or *proton number*) and is denoted by the symbol Z. Obviously, this number also represents the number of electrons associated with that atom. Elements in the *Periodic Table* are simply arranged in ascending

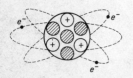

Fig. 20:1 Schematic representation of a lithium atom.

order of atomic number, e.g. hydrogen $Z = 1$, helium $Z = 2$, lithium $Z = 3$, etc.

Since the overall mass of the nucleus is determined by how many neutrons are present within that particular nucleus, the total number of nucleons (protons plus neutrons) is referred to as the **mass number** (or *nucleon number*) A. Hence the actual number of neutrons present in a particular nucleus is given by $(A - Z)$.

Symbolically, nuclei are represented by

$$_Z M^A \qquad (\text{or} \quad {}_Z^A M \quad \text{or} \quad M_Z^A)$$

The subscript gives the atomic number and the superscript the mass number. Using this notation, a nucleus of the element aluminium would be represented by $_{13}Al^{27}$, from which it can be deduced that this particular nucleus contains 13 protons and $27 - 13 = 14$ neutrons.

Rutherford's early experiments which led to the concept of the nuclear atom indicated an upper limit to the nuclear radius of around 10^{-14} m. Considering the lithium isotope $_3Li^7$, and approximating its mass to seven nucleons each of mass 1.67×10^{-27} kg, we see that the density of the nucleus will be of the order of

$$\frac{7 \times 1.67 \times 10^{-27}}{\frac{4}{3}\pi \times (10^{-14})^3} \approx 10^{15} \text{ kg/m}^3 \quad (\text{or } 1\,000\,000\,000 \text{ kg/cm}^3!)$$

This is an extremely high density which is well beyond normal experience, and it emphasises the point made earlier that the atom itself is largely empty space with practically all the mass concentrated in a very small nucleus.

The word **nuclide** is used to refer to a particular species of nucleus with a given value of Z and A. Nuclides with the same value of Z but with different values of A are referred to as **isotopes**. As isotopes have the same value of Z, they also have the same number of electrons and, since it is the electrons which determine the chemical behaviour of the elements, they are indistinguishable by chemical methods. Most elements in the periodic table have one or more stable isotopes. For example, there are three isotopes of neon, $_{10}Ne^{20}$, $_{10}Ne^{21}$ and $_{10}Ne^{22}$, which exist naturally, with relative abundances of 90.8%, 0.3% and 8.9% respectively.

20.2 Measuring the Masses of the Nuclides

The accurate measurement of the actual masses of the various nuclides is essential to any proper understanding of nuclear behaviour. The instrument used to "weigh" the nucleus is called a **mass spectrograph**. Many types of mass spectrograph have been developed since the original work of J. J. Thomson. However they all depend on the same principle of deflecting positive ions in a magnetic field and then inferring the mass of the nucleus from the radius of curvature of the track. (The bending of charged particles in a magnetic field was considered in section 15:6.)

A typical form of mass spectrograph, originated by Bainbridge in 1933, is shown in fig. 20:2. The instrument has three main sections. Firstly there

Fig. 20:2 Bainbridge-type mass spectrograph. The whole instrument operates in a uniform magnetic field of flux density B normal to the diagram.

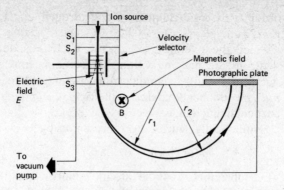

is an ion source which may produce positive ions by means of a simple gas discharge, by heating, or in some cases by irradiation from a radioactive isotope. The ions are then accelerated into the second device which acts as a velocity selector. This consists of a collimating system of slits between which there is an electric field of strength E and also a magnetic field of flux density B acting on the ion beam. This magnetic field is the same field which is used to do the actual bending of the ion trajectories in the third part of the instrument in which the "weighing" is done. The uniform magnetic field is usually produced by having the whole instrument placed between the flat poles of a large electromagnet.

The theory of the instrument is quite simple. In the velocity selector, ions which enter slits S_1 and S_2 can successfully pass through S_3 only if the deflecting effects of the crossed electric and magnetic fields balance, i.e.

only if $\quad Eq = Bqv$

where q is the charge on the ion and v is its velocity. Therefore any ion with a velocity different from that given by

$$v = \frac{E}{B}$$

will be trapped in the system by S_3 and prevented from entering the analysis region. Those that do get through S_3 will have a well-defined velocity.

Once these ions are in the main magnetic field region they are bent into circular trajectories which must satisfy:

$$Bqv = \frac{mv^2}{r} \quad \text{i.e.} \quad m = \frac{Bqr}{v}$$

and substituting for the velocity this becomes

$$m = \frac{B^2 qr}{E}$$

Then, if E, q and B are constant,

$$m \propto r$$

Thus the position of the image of the slit produced on the photographic plate by the particular ion is proportional to its distance from the source slit ($= 2r$). When there is a mixture of ions having different masses present, these will produce a series of lines on the plate corresponding to the different isotopes present. Also the intensity of the fogging produced on the plate will give a measure of the relative abundance of the different isotopes present in the sample. In the most recent form of this equipment the ions are recorded electronically with an ionization chamber, the data is processed by a micro-computer, and the isotopes present with their relative abundance are automatically displayed.

Since 1960 the isotopic masses have been measured relative to the mass of the carbon isotope $_6C^{12}$. This **unified atomic mass unit** u is defined as

$$1\ u = \tfrac{1}{12}\ (\text{mass of }_6C^{12})$$

and the absolute value of this unit then works out to be

$$1\ u = 1.66 \times 10^{-27}\ \text{kg}$$

On this scale the proton and neutron masses become

$$m_p = 1.007\ 277\ u$$
$$m_n = 1.008\ 665\ u$$

Since the mass spectrograph actually measures ions which include a small contribution to the mass from the electrons, another useful mass is that of the neutral hydrogen atom:

$$_1H^1 = 1.007\ 825\ u$$

20.3 The Binding Energy of the Nucleus

In order to understand the behaviour of the nucleus it is necessary to apply one of the most significant results derived by Einstein from the special theory of relativity. Einstein demonstrated that mass and energy are equivalent and are related by the simple equation:

$$E = mc^2$$

where E is the energy (in joules), m the mass (in kilogrammes) and c is the velocity of light in free space (in metres per second).

In effect the equation states that mass is just another form of energy. If, for instance, matter equal to 1 u, i.e. 1.66×10^{-27} kg, were to disappear, in its place would appear a quantity of energy given by

$$E = mc^2 = (1.66 \times 10^{-27})(3 \times 10^8)^2 = 1.494 \times 10^{-10}\ \text{J}$$

On first examination this may not seem a very large amount of energy. The magnitude can perhaps be illustrated by considering the complete annihilation of a small bit of matter weighing approximately 0.01 g say, something like a pinhead. In this case we have

$$E = 0.01 \times 10^{-3}(3 \times 10^8)^2 = 9 \times 10^{11}\ \text{J}$$

This is sufficient energy to place a rocket of many tonnes into orbit in space.

One fact which rapidly became apparent following the arrival of the accurate mass spectrograph was that the mass of any given nuclide was always less than the summed mass of all its constituent nucleons. In other words, when protons and neutrons are combined to create a particular nucleus, mass apparently disappears and, in fact, energy equivalent to the lost mass is released. Conversely, if a given nucleus were to be separated out into its constituent nucleons, energy would have to be supplied equivalent to the missing mass. The missing mass is referred to as the *mass defect* and the energy equivalent of this is called the **binding energy** of the nucleus.

Consider a simple isotope such as $_2He^4$. Since most mass spectrograph data is given in terms of the neutral atom, this isotope which consists of 2p, 2n and 2e may be considered as 2 neutral hydrogen atoms plus 2n. Then, with $m_H = 1.007\,825\,u$ and $m_n = 1.008\,665\,u$, the mass of the isotope should be given by

$$m_{He} = 2m_H + 2m_n = 2 \times (1.007\,825 + 1.008\,665) = 4.032\,980\,u$$

However from measurements made with the mass spectrograph, helium has the value $4.002\,603\,u$. Consequently there is a mass defect for helium of $\Delta m = 0.030\,377\,u$ and the binding energy is therefore given by

$$BE = \Delta mc^2$$
$$= (0.030\,377 \times 1.66 \times 10^{-27}) \times (3 \times 10^8)^2 J$$
$$= 4.538 \times 10^{-12}\,J = 28.36\,MeV$$

This is the energy which would have to be supplied to a helium nucleus in order to break it up into its separate parts of 2p, 2n and 2e.

In comparing the behaviour of different nuclides, what becomes of greater significance than the actual binding energy is the binding energy per nucleon. If this quantity is evaluated for all the known elements, and then plotted on a graph against the mass number of each element, the curve illustrated in fig. 20:3 is obtained. Apart from a few slight

Fig. 20:3 Binding energy per nucleon curve.

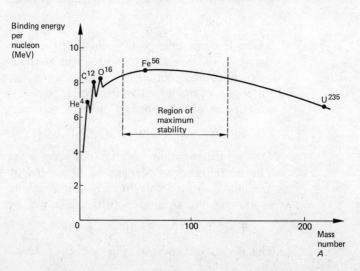

anomalies, e.g. helium (He), carbon (C) and oxygen (O), the variation is smooth, with a maximum value around $A=50$–60 of approximately 8.7 MeV per nucleon. The broad maximum region represents nuclides with the greatest stability since these will require the greatest input of energy in order to break them up into their constituent nucleons. The particular significance of this curve will become apparent in the next chapter when the phenomena of fission and fusion are considered.

20:4 Nuclear Reactions

The transmutation of one element into another was a dream sought by the early alchemists using chemical processes. The actual phenomenon was first demonstrated by Rutherford in 1919. Rutherford and his research group allowed highly energetic alpha-particles from a radioactive isotope to collide with atoms of nitrogen (N) in the air. He was able to detect protons released from the collision and proposed that a nuclear reaction had occurred which could be described by the equation:

$$_7N^{14} + {_2}He^4 = {_8}O^{17} + {_1}H^1$$

Notice particularly, in writing this equation, that the mass numbers (total nucleon number) and atomic numbers (charge on the nucleus), when totalled, are the same on both sides of the equation. In writing similar equations this must always be observed. It simply recognises the fact that charge and nucleons are conserved through the reaction.

The reaction which Rutherford proposed was subsequently verified when it was observed in a cloud chamber (see chapter 22). An example of such a cloud chamber picture is shown in fig. 20:4. The proton is seen to fly away at high velocity, while the relatively massive oxygen isotope recoils, to conserve the momentum in the system, and produces the short heavily ionized track.

In this particular reaction the alpha-particle has to be very energetic to overcome the natural repulsion between it and the nitrogen nucleus and, in addition, to supply the necessary energy to create the increase of mass which is required. When energy must be supplied to the reaction as in this case, the reaction is referred to as *endoergic*.

In this reaction the bombarding particle is an alpha-particle and the released particle is a proton. It is therefore described as an α–p reaction. Such a reaction is then often written in the short form:

$$_7N^{14}(\alpha, p)_8O^{17}$$

Many similar reactions have now been observed, for instance:

$$_{13}Al^{27}(\alpha, p)_{14}Si^{30} \qquad \text{(Si is silicon)}$$

In this case there is a net release of energy and it is therefore referred to as *exoergic*.

Fig. 20:4 (*a*) Cloud chamber photograph of Rutherford (α, p) reaction; (*b*) key to (α, p) reaction in the photograph.

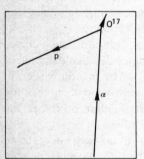

In 1932 Cockroft and Walton demonstrated a nuclear reaction which was produced by particles accelerated up to high energies in a purpose-built machine, instead of relying on alpha-particles from a radioactive source. The instrument used a very high potential to accelerate protons towards a target. Cockroft and Walton first used the instrument to demonstrate a (p, α) reaction in which a lithium atom was transmuted into two helium atoms (alpha-particles), i.e.

$$_3\text{Li}^7 + _1\text{H}^1 = _2\text{He}^4 + _2\text{He}^4$$

The particular machine used by Cockroft and Walton was capable of accelerating particles up to energies of about 0.7 MeV. It was the forerunner of a whole family of **particle accelerators** based on different accelerating principles. These range from straightforward high potentials produced by Van der Graaf type electrostatic generators, to linear accelerators using a travelling electromagnetic wave, and cyclotron and synchro-cyclotron types in which the particles are accelerated by a radio-frequency electromagnetic field and maintained in a circular trajectory by large electromagnets. With such particle accelerators capable of producing high-intensity and high-energy beams up to hundreds of GeV, many different nuclear reactions have now been observed in addition to those already mentioned.

One other type of reaction of importance is that in which the bombarding particle is a neutron, for instance:

$$_5B^{10} + _0n^1 = _3Li^7 + _2He^4 \quad \text{or} \quad _5B^{10}(n, \alpha)_3Li^7$$

(B is boron.) In this reaction, of course, since the neutron has no charge, it cannot be accelerated to high energies by electric fields, and the bombarding particle must be produced from another reaction. This is not a serious problem since, with no charge, it is not repelled by the target nucleus. Consequently, nuclear reactions can be induced with quite low-energy neutrons. One particular reaction induced by a "slow" neutron occurs with the uranium isotope U^{235}. This is a reaction in which a very large quantity of energy is released and it is referred to as fission. We shall be considering this reaction in detail in the next chapter.

Exercises

20.1 Define the terms *nucleon*, *atomic number*, *mass number* and *binding energy*.
Use the Periodic Table of Elements to identify the following isotopes:

$$_{22}X^{50} \quad _{79}X^{197} \quad _{11}X^{22} \quad _{47}X^{107}$$

and in each case list the number of protons and neutrons present in the nucleus.

20.2 Show that 1 atomic mass unit (u) is equivalent to 931 MeV.

20.3 Determine how many units of rest mass must disappear in a nuclear reaction in which 159 MeV of energy is released.

20.4 Calculate the average binding energy per nucleon for the $_6C^{12}$ isotope.

20.5 Complete the following equations, balancing charge and mass number and using the Periodic Table to identify the elements.

$$_4Be^9 + _2He^4 = _6C^{12} + (\ )$$
$$_5B^{10} + _0n^1 = (\ ) + _2He^4$$
$$_{13}Al^{27} + _2He^4 = _0n^1 + (\ )$$
$$_5B^{11} + (\ ) = _6C^{12} + _0n^1$$

20.6 Fill in the missing symbols:

$$_5B^{10}(n, \alpha)? \quad _?Mg^?(\gamma, p)_?^{24} \quad _{17}Cl^{35}(n, p)_?^?$$
$$_?Mn^?(p, n)_?^{55} \quad _{48}Cd^{113}(?, \gamma)_{48}Cd^{114}$$

(Mg magnesium, Cl chlorine, Mn manganese, Cd cadmium)

21 The Unstable Nucleus—Radioactivity

Historically, many of the early discoveries of nuclear behaviour and composition, which were considered in the last chapter, were related to the phenomenon of radioactivity. We have already noted the use of alpha-particles from a radioactive isotope to induce nuclear reactions. In the present chapter the decay of a nucleus by the spontaneous emission of an energetic particle will be considered in some detail, together with the specific properties of the radiations involved.

The initial discovery of radioactivity was made by Becquerel in 1896. The actual discovery was accidental, since he was looking for a phosphorescence effect produced by exposure to sunlight at the time. His experiment consisted of exposing various crystals to sunlight and then wrapping them in photographic paper to see if any fogging occurred after they were removed from the sunlight. One of his samples was a uranium salt which was, by chance, found to have fogged some photographic plates even though it had not previously been exposed to the light. Becquerel followed up his chance discovery and soon showed that the fogging was caused by some form of radiation emanating from the uranium salt.

The new discovery was soon being intensely investigated, notably by Marie and Pierre Curie in 1898. They succeeded in chemically isolating a tiny amount of two new elements, which were primarily responsible for the radiation, from a large quantity of uranium ore. They called the two new elements radium and polonium. Rutherford later showed that the emissions from these elements contained two distinct types of radiation, one of which was easily absorbed and one which was very penetrating. The former was referred to as alpha-radiation and the latter as beta-radiation. Subsequently, a third type of radiation was identified which was even more penetrating than the beta-radiation and this was accordingly called gamma-radiation.

21:1 The Identity of the Radiation

The three types of radiation may be separated out and identified by the simple imaginary experiment illustrated in Fig. 21:1. A source of radiation such as radium is placed at the bottom of a small hole drilled in a large piece of lead, which therefore only allows a narrow beam of radiation to emerge. On entering the electric field, one of the beams is deflected slightly to the negative plate [the alpha-particles], one suffers a large deflection towards the positive plate [the beta-particles], and the remaining beam [gamma-radiation] is undeflected. (N.B. The reason why the experiment is imaginary is that a field sufficiently large to produce a measurable deflection of the alpha-particles would bend the beta-particles onto the positive plate.)

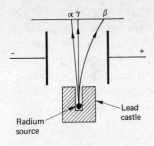

Fig. 21:1 Separation of the radiations.

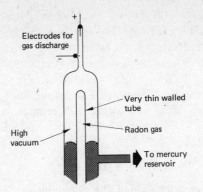

Fig. 21:2 Rutherford and Royds' experiment.

Alpha-particles (α-particles) were suspected of being the nuclei of helium atoms $_2\text{He}^4$. This was eventually proved to be the case by an experiment carried out by Rutherford and Royds in 1909. The essential form of their apparatus is shown in fig. 21:2. Radon gas (an alpha emitter) was trapped in the central thin-walled tube. The walls were sufficiently thin to allow much of the emitted alpha-radiation to pass into the outer region of the double-walled system which had been previously evacuated. In passing through the wall, the alpha-particles picked up a couple of electrons to become neutral helium atoms. The apparatus was then left for about a week, and the helium which had collected in the outer space was compressed up into the small tube at the top of the system, by raising the mercury level. An electric discharge was then passed between the electrodes and the light examined in an optical spectrometer. The spectrum observed was the unique line emission spectrum normally seen from helium gas positively identifying the alpha-particles as doubly ionized helium nuclei. Hence, alpha-radiation consists of particles carrying two positive charges and having a mass of approximately 4 u.

Beta-particles (β-particles) were originally shown to have the same specific charge (e/m) as cathode rays (i.e. electrons) by Becquerel in 1900. Beta-particles therefore carry one unit of negative charge and have a mass of approximately 1/1800 the mass of the proton.

Gamma-radiation (γ-rays), unlike the previous two, does not consist of actual particles of matter. It is identified as very-short-wavelength electromagnetic radiation. On the electromagnetic spectrum it lies in the region beyond X-rays. Since it is electromagnetic radiation, it is undeflected by electric and magnetic fields, as previously noted in fig. 21:1. Also, because of the very short wavelength, gamma-rays are very penetrating and a photon of this wavelength carries a large amount of energy (see chapter 18, fig. 18:5).

21:2 The Process of Nuclear Disintegration

The emission of a particle is a nuclear process which is entirely spontaneous and is not affected by criteria which control chemical processes, such as temperature and pressure. For instance, placing a radioactive isotope in an oven at a temperature of 2000°C in no way alters its decay rate or the energy of the emissions. Chemical reactions are associated with the extra-nuclear electrons and energies involved are measured only in terms of tens of electron-volts, whereas the emission of alpha, beta and gamma radiations is associated with energy changes of millions of electron volts.

When a nucleus emits an alpha-particle, its atomic number Z decreases by 2 and the mass number A decreases by 4. Since both Z and A change, an isotope of a new element is created as a result of the decay. An example of alpha-decay occurs with one of the isotopes of radium (Ra), e.g.

$$_{88}Ra^{226} \rightarrow \,_{86}Rn^{222} + \,_2He^4(\alpha)$$

It is the custom to call the decaying nucleus the *parent* and the newly created nucleus the *daughter*. In this particular example, the daughter nucleus is the gas radon (Rn) which is itself an alpha-particle emitter:

$$_{86}Rn^{222} \rightarrow \,_{84}Po^{218} + \,_2He^4(\alpha)$$

In beta-particle emission the nuclear charge is increased by one but the mass number remains unchanged because the process results from a change of identity of one of the nucleons within the nucleus. In fact a neutron changes to a proton and an electron is emitted, hence Z goes to $Z+1$, and A remains the same. An example of this type of decay is

$$_{82}Pb^{214} \rightarrow \,_{83}Bi^{214} + \,_{-1}e^0 \qquad \text{(Pb lead, Bi bismuth)}$$

As we have seen with radon, the isotopes created as a result of a decay process are, themselves, frequently unstable and decay yet again. In some cases a whole series of decays can occur until eventually a stable isotope is produced. A well-known example of this is the uranium series which starts with the isotope U^{238} and ends with the stable isotope of lead Pb^{206}. This particular series is illustrated in fig. 21:3. Other known series begin with isotopes of actinium and thorium.

The emission of gamma-radiation is found to be a secondary process usually associated with the emission of alpha or beta particles. Often,

Fig. 21:3 The Uranium series.

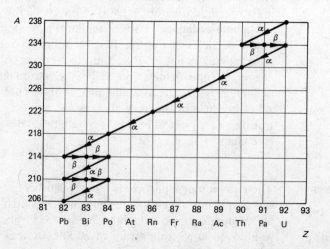

following the emission of a particle, the daughter nucleus is frequently left in an excited state. The nucleus, like the atom, can only exist at certain allowed energy levels and, like the atom, it removes its excess energy and returns to the ground state by emitting photons of electromagnetic energy. The photon energy, as before, is related precisely to the change in energy level involved. The only difference between the nuclear and atomic cases is one of magnitude. The energy changes involved in the nucleus are vastly greater than those in the atom. Instead of photon energies measured in tens of electron-volts, the energies of gamma-photons are of the order of MeV, e.g. the cobalt Co^{60} isotope emits two gamma-photons of energies 1.33 and 1.17 MeV. Obviously, because of the nature and origin of gamma-radiation, no change of atomic or mass number occurs as a result of a gamma emission.

21:3 The Properties of Alpha, Beta and Gamma Radiation

1 **Alpha-particles** are very effective at producing fluorescence in materials such as zinc sulphide (the same type of material with which CRO and TV screens are coated) and they also readily fog photographic paper (i.e. they produce blackening on a film after development, just as though the film had been exposed to light). They also ionize gases very strongly. This occurs because the alpha-particle is a relatively heavy particle and is travelling at high velocity. When the particle passes close to a gas atom, the Coulomb attraction between it and the negative electrons of the atom is sufficient to strip off one or more of those electrons, thus creating free electrons and positive ions in the gas. The alpha-particle consequently leaves a trail of ion pairs in its wake as it passes through the gas. On average, the alpha produces approximately 10^5 ion pairs per centimetre of track in air at S.T.P. Each ionizing event requires work to be done and, since this work is derived from the alpha-particle's original kinetic energy, the range of the particle is limited. At the end of the range when the alpha is finally stationary, it captures two electrons and becomes a neutral helium atom. At this point the alpha is said to have been *absorbed*. It is found that this absorption occurs at a definite range for alphas emitted by a given nuclide. For gases this range would typically be of the order of a few centimetres, while for a solid material such as aluminium it is only 0.1 mm. In fact alpha-particles are quite easily stopped by a sheet of paper.

The fact that there is a definite range indicates that the alpha-particles were all emitted with a precise single energy, and this energy can be related to the change of energy (including mass) between the parent and daughter nuclei. In some cases, nuclides emit groups of mono-energetic alpha-particles, each group having a precise energy, and in this situation each group energy can be related to allowed nuclear quantized energy levels with which the daughter nucleus can exist.

2 **Beta-particles** are also found to cause fluorescence and they fog photographic material. They also ionize gases but much less strongly than alpha-particles, producing about 100 ion pairs per centimetre of track in

air at S.T.P. They can penetrate matter more easily than alphas, up to several metres in air and one or two centimetres in aluminium. However, unlike alpha-particles, they have no clearly defined range in air, indicating that they are emitted with a wide spectrum of energies from zero up to some maximum value.

The observed range of energies possessed by the beta-particles did for some time prove quite a problem for those physicists studying the decay process. It appeared that the law of energy conservation was being violated. The difficulty was resolved following a suggestion by Pauli that a second particle, which he called the **neutrino** (ν), was emitted at the same time as the beta-particle. The disintegration energy was then shared between the two particles such that

$$\Delta E_{\text{nucleus}} = E_{\text{beta}} + E_{\text{neutrino}}$$

as illustrated in fig. 21:4.

The neutrino is an extremely difficult particle to detect; in fact for a long period Pauli's explanation was not accepted because no one could detect its presence! The particle was eventually detected by Cowan and Reines in 1953 and it has been the subject of much study since.

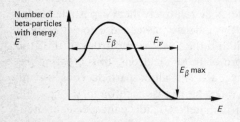

Fig. 21:4 Energy spectrum of beta particles.

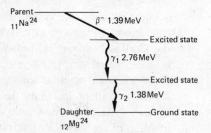

Fig. 21:5 Gamma line emissions from $_{12}\text{Mg}^{24}$.

3 Gamma-radiation is clearly different from the previous two types of emission in that it is electromagnetic radiation. It does, however, still produce fluorescence in certain materials, e.g. barium platino-cyanide, and it also causes fogging of photographic material. The degree of ionization produced in gases is considerably less, giving only a few ion pairs per centimetre of track in air at S.T.P. As a result of this and also the very short wavelength of the radiation, it is extremely penetrating and is able to pass through several centimetres of lead. The emission of gamma-radiation occurs when the daughter nucleus of the decay is formed in an excited state. The gamma photons emitted represent the removal of the excess energy as the new nucleus returns to its lowest ground state energy. The gamma-photons are of specific energy related to the allowed energy levels of that particular nuclide. The process is illustrated in fig. 21:5 for the unstable sodium isotope $_{11}\text{Na}^{24}$ which decays by beta-emission to an excited state of the daughter isotope $_{12}\text{Mg}^{24}$ (magnesium). Two gamma-photons are observed, of energies 2.76 MeV and 1.38 MeV.

21:4 The Laws of Radioactive Decay

The process of radioactive decay is observed to be a perfectly random phenomenon, governed by the laws of chance. Each unstable nucleus has a definite probability of disintegrating or not disintegrating in any given interval of time, and this probability is constant for any particular nuclide and is also independent of any external factors, e.g. temperature.

Suppose that N atoms of a certain nuclide were being observed with suitable equipment which enabled each of the disintegrations to be counted. The number of original nuclei seen to decay would be proportional to both the number of atoms originally present and to the period for which they were observed. This can be expressed mathematically as

$$\Delta N \propto N \, \Delta t$$

where ΔN represents the change in the number of original atoms and Δt the interval of time for which they were observed.

To convert this proportionality to an equality, a constant is introduced, in the usual way, called the **decay constant** λ. This constant is a measure of the probability of decay for that particular species of nuclide. Thus

$$\Delta N = -\lambda N \, \Delta t$$

The negative sign is introduced to recognise the fact that the number of original atoms gets smaller as the time interval increases.

Then re-arranging:

$$\frac{\Delta N}{\Delta t} = -\lambda N$$

and, using the notation of calculus, this becomes

$$\frac{\mathrm{d}N}{\mathrm{d}t} = -\lambda N \qquad (21:1)$$

where $\mathrm{d}N/\mathrm{d}t$ represents the rate of change of the number of original nuclei or, in other words, the number observed to decay per unit time.

If no new atoms are being added to the system, then the number which have decayed after a time t has elapsed from an initial time $t = 0$ is obtained by the mathematical process of integration, which gives the **law of radioactive decay**:

$$N_t = N_0 e^{-\lambda t} \qquad (21:2)$$

where N_t is the number of atoms present after time t and N_0 is the original number at $t = 0$.

Hence from equation (21:2) the number of atoms present is seen to change exponentially with time. The shape of the curve produced when the value N_t is plotted against time is shown in fig. 21:6.

In order to obtain an appreciation of how fast a given nuclide is decaying, a quantity called the **half-life** $T_{1/2}$ is introduced. The half-life is defined as the time required for half of the original number of nuclei to decay. Thus in a time equal to one half-life,

$$N_t = \tfrac{1}{2} N_0$$

If this is substituted into equation (21:2), then

$$\tfrac{1}{2}N_0 = N_0 e^{-\lambda T_{1/2}} \quad \text{i.e.} \quad \tfrac{1}{2} = e^{-\lambda T_{1/2}} \quad \text{giving} \quad 2 = e^{\lambda T_{1/2}}$$

Therefore $\log_e 2 = \lambda T_{1/2}$ so that $T_{1/2} = \dfrac{0\cdot693}{\lambda}$

Obviously in each interval of time equal to one half-life, the number present decreases by half as illustrated in fig. 21:6. Half-lives can vary over very wide limits extending from 10^{-7} sec to 10^{10} years.

Fig. 21:6 The exponential decay curve.

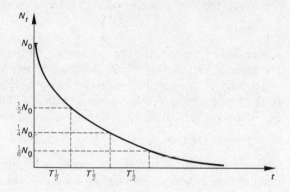

Experimental work on radioactivity is normally concerned with counting the number of disintegrations occurring per unit time, i.e. dN/dt. This quantity is referred to as the **activity** A and has units called Becquerel. One BECQUEREL (Bq) is equivalent to one disintegration per second.

$$\text{Since} \quad A = \frac{dN}{dt} = -\lambda N$$

then $\quad A \propto N$

Thus the activity of any sample of radioactive isotope will change in exactly the same way as the number of atoms present and therefore we can also write

$$A_t = A_0 e^{-\lambda t}$$

Example 21:1

A sample of the nuclide $_{11}Na^{25}$ (sodium) is observed to decay emitting beta-particles ($_{-1}e^0$). The sample weighs 0.5 mg and the half-life of the isotope is 62 sec. Determine a) the identity of the daughter nuclide, b) the number of atoms of $_{11}Na^{25}$ initially present, and c) the activity of the sample after one hour.

Solution
a) The decay can be represented by

$$_{11}Na^{25} \rightarrow {}_{12}X^{25} + {}_{-1}e^0$$

since charge and nucleon number must balance on both sides of the relation. By reference to a Periodic Table the daughter nuclide is identified as an isotope of magnesium $_{12}Mg^{25}$.

b) The relative atomic mass of sodium is 25 (obtained approximately by equating it to the mass number). Hence using Avogadro's number, the number of sodium atoms initially present is given by

$$N_0 = \frac{6.02 \times 10^{23} \times 0.5 \times 10^{-3}}{25} = 1.2 \times 10^{19} \text{ atoms}$$

c) We have that the activity of any sample is given by

$$A = \lambda N \quad \text{(ignoring minus sign)}$$

Hence, initially, $A_0 = \lambda N_0$

But the half-life is $T_{1/2} = \dfrac{0 \cdot 693}{\lambda}$ $\quad \therefore \quad \lambda = \dfrac{0 \cdot 693}{T_{1/2}}$

Then on substituting

$$A_0 = \frac{0.693}{T_{1/2}} N_0 = \frac{0.693}{62} \times 1.2 \times 10^{19} = 1.34 \times 10^{17} \text{ Bq}$$

Finally, using the exponential law:

$$A_t = A_0 e^{-\lambda t} = 1.34 \times 10^{17} \times e^{-0.693 \times 3600/62} = 0.45 \text{ Bq}$$

21:5 Applications of Radioactivity

With the advent of nuclear reactors it is now possible to produce unstable isotopes of virtually all known elements by exposure to the high neutron flux in the reactor. These artificially produced isotopes, together with the naturally occurring radioactive isotopes, can be put to various uses in industry, medicine and biology, geology and archaeology.

Industrial uses include automatic measurement of the thickness of various products. Radiation is absorbed in a definite way when passing through different materials and, by monitoring the intensity after passage through the material, a precise indication of the thickness can be determined. A typical application might be automatic control of a roller press to produce sheet steel of a given thickness in a continuous process. The signal generated by the detector is analysed by a micro-computer which then automatically sets the machinery to give the required results.

Within the field of biology and medicine, two main applications arise. One is radiation therapy where, typically, gamma-radiation is used to kill cancer cells within a tumour. Normally, many beams are used which all focus onto the location of the tumour from different directions, thus concentrating the effect of radiation at the site of the tumour while minimizing the damage to good cells in the surrounding tissue. Gamma-radiation is also used for general sterilization of food, dressings and equipment.

Another important application of specific isotopes is their use as tracers in both biological research and medicine. Isotopes of different elements are related to different metabolic pathways and cellular processes in very specific ways. A typical example might be the use of the iodine isotope I^{131}, which, when injected into the body, is taken up exclusively by the thyroid gland. Hence by monitoring the presence of activity from the I^{131} (which has a half-life of 8 days) in the body and at the site of the gland, an assessment of the gland's performance can be made. In addition, the presence of the active iodine in the gland can be used to kill off some of the cells within the gland and hence reduce its function in cases of overactive thyroid.

There are many elements which can be "labelled" by inclusion of a small proportion of radioactive isotopes and these labelled compounds can be used to monitor all sorts of biological and physical processes. One such example is in the study of wear in moving parts of a piece of machinery. An artificially produced isotope Fe^{59} is frequently used for this purpose. The metal of, say, a bearing is made up incorporating the isotope and, by examining the amount of active material which can be detected in the oil circulating through the bearing, an estimate of the rate of wear can be deduced.

Geology makes use of many naturally occurring isotopes to provide an estimate of the age of rocks and indeed the age of the earth itself. We saw in fig. 21:3 that the isotope $U^{238}(T_{1/2} = 4.5 \times 10^9$ years) eventually decays to a stable isotope of lead Pb^{206}. By measuring the relative proportions of Pb^{206} and U^{238} in a uranium-bearing rock, a calculation of the age of the rock can be made. Other useful isotopes for this purpose are K^{40} which decays by electron capture to an isotope of argon, and rubidium Rb^{87} which decays by normal beta-emission to an isotope of strontium. Both these have suitably long half-lives to permit the age of the earth to be estimated. Consideration of these various decays leads to a conclusion that the age of the earth is approximately 4.8×10^9 years.

A much shorter half-life isotope if of great importance in the dating of archaeological material of relatively recent history. The method is known as Carbon-14 dating since it makes use of the isotope C^{14}. This isotope is continuously generated in the upper atmosphere by cosmic ray activity in the neutron reaction:

$$_7N^{14} + {_0}n^1 \rightarrow {_6}C^{14} + {_1}p^1$$

whereupon the C^{14} isotope decays:

$$_6C^{14} \rightarrow {_7}N^{14} + {_{-1}}e^0$$

with a half-life of 5570 years approximately. Normally, in the atmosphere and in living cells, the amount of C^{14} present (approximately 1 in every 10^{12} of normal C^{12}) is in equilibrium. However, when the animal or plant dies, the exchange ceases and the C^{14} content progressively decays away from then on. By measuring the C^{14}/C^{12} ratio present in the dead material, e.g. a piece of wood, the age can be estimated if the equilibrium ratio is assumed to have remained more or less constant.

21.6 Fission

When the binding energy per nucleon curve is plotted across all elements, a surprisingly regular pattern results, except for a few small peaks as has been noted already in chapter 20 (fig. 20:3). The maximum binding energy per nucleon occurs around $A = 60$ and hence these nuclides are the most stable, requiring the greatest amount of energy to disrupt them. Thus, if a heavy nucleus such as $_{92}U^{235}$ were to split into two lighter ones

Fig. 21:7 The fission process.

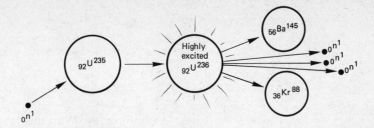

with A around 100, a large amount of energy would be released since the binding energy per nucleon is increased. This process is called **nuclear fission**.

The fission process occurs particularly easily in the case of the isotope U^{235} which requires only the absorption of a low-energy neutron (sometimes called a thermal neutron). The process is illustrated in fig. 21:7. The absorption of the slow neutron produces an excited state of the isotope U^{236} (it acquires 6.8 MeV from the binding energy of the extra neutron). It happens that extra energy of this magnitude is enough to set the nucleus into violent oscillation, rather like a liquid drop, which is of sufficiently great amplitude for it to break up. Each fission is different in terms of the fission products which result. The example shown gives isotopes of barium and krypton, while another might give isotopes of xenon and strontium and so on. Notice also that the fission process results in the emission of further neutrons which can vary in number between 0 and 9 and, on average, number approximately 2.5 per fission.

The reaction may be written:

$$_{92}U^{235}(n, \gamma)_{92}U^{236*}$$

followed by

$$_{92}U^{236*} \rightarrow {}_{56}Ba^{145} + {}_{36}Kr^{88} + 3\,_0n^1$$

or $\quad _{92}U^{236*} \rightarrow {}_{54}Xe^{140} + {}_{38}Sr^{94} + 2\,_0n^1$

The * indicates an excited state.

The fission process releases a great deal of energy which is mainly in the form of kinetic energy of the fission fragments and released neutrons. Consider

Binding energy/nucleon for $A = 235$ is approximately 7.6 MeV.

Binding energy/nucleon for $A = 120$ is approximately 8.5 MeV.

Hence the change in BE/nucleon is 0.9 MeV. Consequently the energy released is approximately 235×0.9, i.e. about 200 MeV per fission. This should be compared with figures of eV or tens of eV for chemical reactions. In joules this is equivalent to

$$200 \times 10^6 \times 1.6 \times 10^{-19} = 3.2 \times 10^{-11} \text{ J}$$

which perhaps does not seem particularly great. However, this, it must be remembered, is the energy for just one fission. If 1 g of U^{235} could be caused to undergo fission, the associated energy release would be given by

$$\frac{N_A}{235} \times 3.2 \times 10^{-11} \, \text{J}$$

where N_A is Avogadro's Number. Hence the total energy is

$$\frac{6.02 \times 10^{23} \times 3.2 \times 10^{-11}}{235} = 8.2 \times 10^{10} \, \text{J}$$

This is a very great deal of energy. With this amount of energy it would be possible to boil 200 000 kg (or 52 800 gallons) of water, or if released all at once in approximately one micro-second, an explosion equivalent to 20 tons of TNT.

The possibility of using a fissionable material for energy production is crucially dependent on the neutrons released as part of the fission process. If at least one neutron goes on to induce another fission in a nearby nucleus, a **chain reaction** can be sustained. When the chain reaction is uncontrolled, a nuclear fission bomb results. If the neutron production is controlled by the use of moderators and absorbers, a **nuclear reactor** results which can be used to generate electrical power.

A much simplified nuclear reactor is illustrated diagrammatically in fig. 21:8. The fuel elements are tubes of high-melting-point metal which contain a mixture of U^{235} and U^{238} isotopes. *Moderators*, which surround the fuel, are made of light elements such as graphite and these slow down most of the high-speed neutrons by means of a billiard ball type of collision. The kinetic energy of the neutron is then exchanged with the atoms of the moderator. Some of the high-speed neutrons react with U^{238} before being slowed down and, as a result, produce plutonium which itself can be used as a nuclear fuel. The neutrons slowed down in the moderator go on to produce fissions in more U^{235}, which then releases more neutrons to sustain the chain reaction.

The overall control of the reactor is maintained by the use of *control rods* constructed from a neutron-absorbing material such as cadmium. These act like blotting paper absorbing out the neutron flux and thus controlling the rate at which the chain reaction can proceed and consequently the rate of energy release. Energy is then removed from the core by primary heat exchangers through which either gas, water under pressure, or liquid metal circulates. This heat is then used to produce steam which drives the turbine generators.

The whole reactor must be surrounded by very thick shielding to protect the operators from the high levels of radiation, mainly neutrons and gamma rays. Very elaborate and strong structures must be used to house the whole complex in order to protect the environment in the event of accidents and malfunctions. Other problems associated with the generation of power from the fission process are limitations to the supplies of natural uranium and, very importantly, the subsequent disposal of the radioactive waste which results from the fission products. Despite these obvious problems, fission power may be essential to bridge the gap between the exhaustion of the fossil fuels and the commercial development of alternatives such as fusion which is briefly considered next.

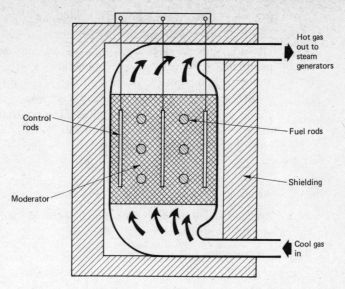

Fig. 21:8 A simplified nuclear reactor.

Hot gas out to steam generators

Control rods

Fuel rods

Shielding

Moderator

Cool gas in

21:7   Fusion

The process of **nuclear fusion** is the converse of nuclear fission. In fusion, light elements are combined to produce heavier elements which again means that a move towards the maximum of the binding energy per nucleon occurs (see fig. 20:3). Hence, as in fission, this must result in the release of energy. In particular, owing to the exceptionally high binding energy of the helium isotope $_2\text{He}^4$, the combination of hydrogen nuclei (protons) to produce $_2\text{He}^4$ results in a very large release of energy. The particles involved have to be brought together at very high collision velocities which are normally achieved in a gas only at extremely high temperatures; so high, in fact, that all the gas atoms are stripped of their electrons and in this situation the gas is normally referred to as a **plasma**. Since the fusion process requires high temperatures for it to take place, it is called a **thermonuclear reaction**. The "burning" of hydrogen to produce helium is the reaction which powers the sun and stars; the initial energy required to generate the necessary temperatures (many millions of degrees celsius) comes from potential energy released as the star collapses in its own gravitational field.

The fusion reaction does not occur directly by collision between four protons but takes place as a cycle of reactions; the net effect being the conversion of four protons into a helium nucleus. A total energy of approximately 27 MeV is released. This may seem small compared to the 200 MeV per fission but it must be remembered that the energy release per nucleon for the fusion process is around 7 MeV compared to 0.9 MeV for fission. Hence, for equal masses of hydrogen and uranium the energy released from the fusion of hydrogen is by far the greater.

Uncontrolled fusion has already been achieved in the form of the hydrogen bomb in which the temperatures necessary for the thermonuclear reaction are generated by a primary fission bomb. However, research continues to achieve a properly controlled fusion reaction which can be used for power production. The advantages are immense, particularly the fact that there is an inexhaustible supply of fuel in the form of water, and also the process does not produce any radioactive waste. The essential problems, however, are producing the appropriately high temperatures and then being able to maintain these temperatures within the plasma for a sufficiently long time, while at the same time maintaining the plasma stable in a "magnetic bottle". Research continues and it is quite possible that power generation from a fusion reactor will be achieved before the end of the century.

Exercises

Avogadro's number $= 6.02 \times 10^{23}$ per mole

21.1 Briefly describe the properties of and identify the ionizing radiations emitted from unstable isotopes.

21.2 Complete the following:

$$_5B^{12} \rightarrow (\quad) + \beta$$

$$_{(\quad)}Bi^{(\quad)} \rightarrow {}_{81}Tl^{208} + \alpha$$

$$_{19}K^{40} \rightarrow {}_{20}Ca^{40} + (\quad)$$

21.3 The isotope $_{37}Rb^{90}$ decays into $_{38}Sr^{90}$. State what kind of particle is emitted.

21.4 The isotope $_{92}U^{239}$ undergoes two successive beta decays. Name the resulting isotope.

21.5 The following counts were obtained from a sample of a radioactive isotope using a Geiger-Muller counter and a scaler.

Time (min.)	Number of counts
0–5	632
5–10	233
10–15	85
15–20	32
20–25	11

By plotting an appropriate graph obtain a value for the half-life of the sample.

21.6 A sample of water contains the isotope tritium $(_1H^3)$ which has a half-life of 12.5 years against beta-decay. Determine the fraction of the original pure tritium which will remain after 25 years.

21.7 The half-life of $_{11}Na^{24}$ is 15 hours. Determine how long it takes for 95% of a sample of this isotope to decay.

21.8 A quantity of the unstable isotope $_{15}P^{32}$ has an activity of 4×10^{10} Bq. If the half-life is 14.3 days, calculate the mass of the sample.

21.9 Measurements show that a 1.0 mg sample of uranium-238 emits 740 alpha-particles per minute. If the atomic mass is taken as 238, determine a value for the half-life of the isotope.

22 Radiation Detectors and Safety

In this final chapter we will briefly review the main methods of detection of all forms of ionizing radiation and consider, also, the essential aspects of safety involved with the handling and use of radioactive isotopes. Radiation detectors now in common use fall into two broad categories, which may be separately described as **electronic particle detectors** and **track detectors**. The particle detectors will be considered first.

22:1 The Electroscope

The **electroscope** is the simplest of all detectors and was used extensively by the early researchers. The principle of operation depends on the phenomenon of gas ionization produced by the alpha, beta and gamma radiations, as illustrated in fig. 22:1. Ion pairs are produced in the air near the charged leaf. Those of opposite sign are attracted to it, resulting in the discharge of the electroscope and consequent fall of the gold leaf. If enough ionising events occur, the electroscope will eventually be totally discharged. The instrument is not much in use today but does however survive in the form of a **dosimeter**. Here the device is miniaturized and constructed to resemble a pen which is worn in the pocket of the person likely to be exposed to any ionizing radiation. The rate of fall of the tiny quartz fibre (which is equivalent to the gold leaf) is monitored and this gives a measure of the intensity of the radiation suffered by the wearer. The significance of this will be considered later under safety.

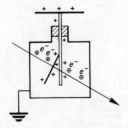

Fig. 22:1 A gold leaf electroscope. The electrons created by the radiation are attracted to the leaf and the positive ions are repelled to earth.

22:2 The Geiger–Muller Counter

This device, like the electroscope, depends on gas ionization. An electric field applied between the central wire and the outer case sweeps the ions to the appropriately signed electrode and the current pulse produced is detected by a sensitive amplifier. The basic form is shown in fig. 22:2.

235

Fig. 22:2 The Geiger-Muller counter.

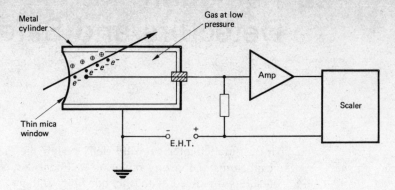

When the potential difference between the wire and outer case is relatively small, an ionization current is produced proportional to the initial number of ion pairs created by the passage of the radiation through the chamber. As the potential is increased further, the electric field near the thin central wire becomes sufficiently large to accelerate the released electrons up to velocities which will generate further ionization by collision.

Eventually an avalanche of ionization is produced involving all the gas atoms contained within the tube. The output pulse produced now is totally independent of the number of ion pairs initially created, and in fact the whole process may be triggered by a single ion pair. A tube operated in this manner is referred to as a **Geiger–Muller counter** after its inventors, and is perhaps the best known of all radiation detectors.

The characteristic behaviour of the tube is shown in fig. 22:3. Geiger–Muller tubes are normally operated with an applied potential somewhere near the middle of the so-called plateau region (typically several hundred volts). This plateau should ideally be horizontal but in actual tubes it has a small slope and the value of the slope is usually quoted by the manufacturer.

The avalanche discharge in the high electric field near the central wire mostly involves the movement of electrons, but of course for every electron present there is also a positive ion. The positive ions being much heavier than the electrons accelerate much less rapidly and, sometime

Fig. 22:3 The different counting regions of the basic gas ionization counter.

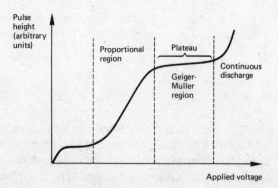

after the initial discharge has gone through, many positive ions are still travelling toward the negative electrode. They eventually strike it with sufficient kinetic energy to eject more electrons, which in turn will initiate a further discharge. To prevent this happening, a quenching agent is included with the operating gas, typically about 0.1% of a halogen such as bromine. The effect of the quenching agent is to absorb most of the positive ion's kinetic energy, thus preventing the secondary emission of further electrons, and hence also continuous discharge.

When in use, it is the negative pulse, created by the electrons following the passage of some ionizing radiation through the chamber, which is counted by the monitoring electronic equipment. While the slow positive ions are moving to the cathode, the Geiger–Muller tube is momentarily inactive and, when counting very rapid events, allowance for this so-called *dead time* must be made.

The Geiger–Muller counter is extremely efficient at detecting beta-particles and it is therefore mainly used for this purpose. However, it will also detect alpha-paticles provided the end window can be made thin enough not to absorb the alphas before they can enter the active gas region. The efficiency in terms of gamma rays is very low with perhaps less than one per cent of the gamma-photons being recorded. This occurs because the gamma rays produce little ionization in the gas within the tube and, even when a gamma ray is detected, this is frequently the result of secondary electrons ejected by the gamma-photon from the wall of the tube.

By filling the Geiger–Muller tube with boron trifluoride gas, it can be made to detect neutrons, which are normally difficult particles to detect since, having no charge, they produce little ionization in their passage through gases. The neutrons interact with the boron:

$$_5B^{10} + _0n^1 \rightarrow _2He^4 + _3Li^7$$

and the alpha-particle created within the tube produces ion pairs which then trigger the tube in the usual way.

22:3 The Scintillation Counter

Although the Geiger–Muller counter is perhaps the most widely known, the **scintillation counter** is the most widely used. The Geiger–Muller has two major problems: the dead time which severely limits the rate of counting and requires lost counts to be estimated, and the poor efficiency for gamma rays.

The scintillation counter, which is illustrated in fig. 22:4, consists of two parts: (i) a scintillating material which emits photons of light when struck by ionizing radiation and (ii) a photo-multiplier which can detect these photons and converts the energy into a measurable pulse of electrons. Scintillating materials can be in solid, liquid or gaseous form. The most common solid used is sodium iodide (NaI) which has been doped with a small amount of thallium (Tl). When a gamma ray or charged particle causes fluorescence in the scintillator, the light photons emitted are

Fig. 22:4 Schematic diagram of a scintillation counter showing the scintillating crystal and photo-multiplier tube.

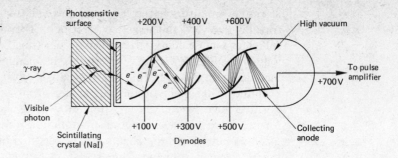

caused to strike a photo-sensitive surface by a mirror and/or light pipe system. On striking the photo-sensitive surface, one or more electrons are ejected which are then attracted and accelerated to the next "dynode" which is maintained at a higher positive potential. Here, each electron ejects more electrons and the process is repeated from dynode to dynode, resulting in a sizeable pulse of current which can easily be detected by the usual electronic instrumentation.

By comparison with the gaseous detecting medium of the Geiger–Muller counter, solid crystal scintillation counters have the advantage of a relatively high density material with a corresponding high electron density. This gives the counter a very high efficiency in the detection of gamma-radiation since the probability of absorption is very high. Scintillation counters can be used for the detection of all forms of radiation and, in addition, they are very fast, and capable of resolving events closely spaced in time since there is no dead time like the Geiger–Muller counter. A scintillation counter can record events up to 10^9 per sec whereas a Geiger–Muller has an upper limit of 5×10^3 per sec. Another great advantage occurs when liquid scintillators are used. These can be mixed with a radioactive sample, thus permitting the detection of very-low-energy radiation since there is no barrier such as a window between the sample and the scintillator.

22:4 The Wilson Cloud Chamber

An extremely important form of track detector was devised by C. T. R. Wilson in 1911. It is based on the same principle which produces fog in certain weather conditions. A fog consists of tiny water droplets which have condensed out of the air when it has become supersaturated, perhaps because of a drop in the temperature. Small nuclei such as dust particles form points at which the condensation process can begin.

This same process goes on inside the **cloud chamber**. The form of the device is illustrated in fig. 22:5. The air inside the chamber is saturated with water or possibly alcohol vapour. When the piston is suddenly withdrawn, the pressure and temperature inside the chamber are reduced, leaving the air in a supersaturated state. The air inside the chamber is kept scrupulously clean and there are, therefore, no dust particles on which condensation can begin and so, in the absence of any other nuclei, condensation begins on any ions present in the air. Hence, if the chamber

Fig. 22:5 The principle of the Wilson cloud chamber.

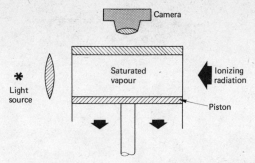

is expanded just after some ionizing radiation has passed through, the track of the particle will be revealed as a line of tiny liquid droplets. In research work involved with cosmic rays and high-energy particles from accelerators, the expansion of the chamber and the photographic recording of the tracks are normally synchronized by picking up the incoming radiation with Geiger–Muller counters arranged in coincidence. If a particle passes though the top tube and also out through the bottom, the system is triggered and a picture is taken. Examples of cloud chamber photographs are shown in figs. 20:4 and 22:6.

A development of the cloud chamber is the **bubble chamber** which operates on somewhat similar principles. Here a liquid is kept under pressure and well above its normal boiling point. When the pressure is momentarily removed, the liquid becomes superheated and boiling begins. In the absence of any dust nuclei or rough edges in the chamber, vapour bubbles initiate on any ions present. Hence, an ionizing particle which passes through the chamber at the right time will leave behind a track of tiny vapour bubbles. The advantage of the bubble chamber is the higher-density medium which means that energetic particles are likely to be stopped within the volume of the chamber. If the active liquid is hydrogen this also enables proton interactions to be studied.

22:5 Other Forms of Radiation Detector

Several other forms of radiation detector are currently in use. Among the particle detectors it is worth mentioning the **solid state detector**. This consists simply of a p–n junction diode. Ionizing radiation which passes through the depletion layer will create extra charge carriers, making the junction more conducting, and the resulting pulse of current can be detected by the usual scalar or rate-meter. The device can detect all three emissions from radioactive isotopes.

The **photographic emulsion** method lies midway between the classifications of particle detector and track detector, Photographic emulsions were one of the earliest methods of detecting the presence of ionizing radiation. In its simplest form the emulsion is still used for this purpose, particularly for detecting X-rays and for use as a radiation dosimeter for people who are working in a hostile environment in terms of ionizing radiation. A small piece of film is worn as a badge during working hours and the intensity of the fogging produced on development can be related to the level of exposure.

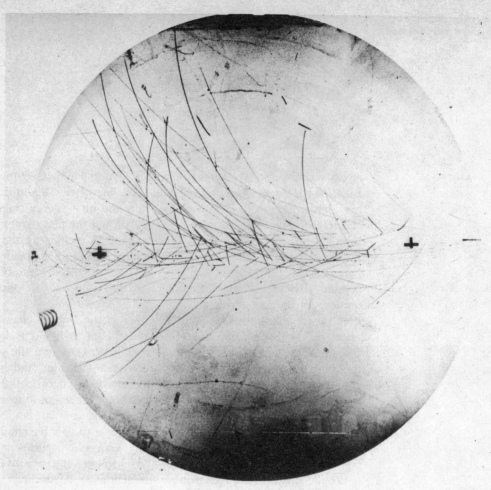

Fig. 22:6 Tracks of recoil protons in a hydrogen-filled cloud chamber traversed by a beam of neutrons. The tracks are curved due to a magnetic field applied at right angles to the plane of the photograph.

22:6 Handling Sources of Ionizing Radiation

All laboratories, industrial sites, factories, etc. where sources of ionizing radiation are in use, and also the actual sources themselves, must be clearly marked with the international hazard symbol (see fig. 22:7).

The storage facilities for radioactive material must also be provided with adequate shielding. As might be expected, the greatest problem is the gamma-radiation. To produce an attenuation of 1000 for gamma-radiation of about 1 MeV, shielding of approximately 70 cm of concrete, 20 cm of steel or 10 cm of lead is required.

Persons working with radioactive material and other sources of radiation must wear appropriate personal-monitoring equipment such as a film badge or electroscope-type dosimeter, to determine accumulated dosage. In particular, radioactive substances should not come into contact with the skin, and disposable gloves and protective clothing should always be worn. Sealed sources should be manipulated using remote handling devices, e.g. tongs or mechanical arms, and open-source operations must only be performed in a fume cupboard or manipulator glove box. Finally, common sense demands that on no account should food or drink be taken in any situation where radioactive materials are present, since ingested material is far more dangerous than external exposure.

Fig. 22:7 The international radiation hazard symbol.

Answers to Numerical Exercises

Chapter 2   Forces and motion
1. 1.25 m/s^2, 90 m, 24 s
2. 30 m/s, 10 s
4. 12.5 m/s, 21.65 m/s
5. 19.6 N
6. 0.96 m/s^2 at $38.6°$ to the 15 N force
7. Forces are tension, weight, normal reaction, and friction down the plane
8. 13.6 N, 5.7 N
9. 60 kN
10. 13.3 rev/s
11. 5.56 rad/s^2
12. 0.167 rad/s, 2.5 m/s^2
13. 0.224 m/s^2
14. 27.3 mm

Chapter 3   Moments and equilibrium
1. $2.21 \times 10^5 \text{ N m}$
2. 415.8 N
3. 2.09 rad/s^2, 95.5 kg m^2
4. 200 s
5. 0.75 m
6. 650 N at A, 850 N at B
7. 57.1 cm
8. 11.8 N, 15.7 N
9. 11.3 kN, 22.6 kN (compression)
10. $30.7°$

Chapter 4   Momentum and energy
1. 3.46 J, 3.72 m/s
2. $4.55 \times 10^{-19} \text{ J}$, $9.1 \times 10^{-18} \text{ N}$
3. 31.9 m, 6.75 J
4. 221 MW
5. 0.75 N s, 30 N
7. 1.5 m/s in original direction, no change in total kinetic energy
8. 0.1 m/s, 200 g
10. 44.5 kJ
11. 6900 N s
12. 875 J, 10 m/s
13. 2 kW
15. 15.7 J
16. 8.63 J, 1.92 J s

Chapter 5   Vibrational motion
1. 2.06 Hz
2. 8.34 m/s^2
3. 1.9 s
4. 0.3 m, 18.8 m/s
5. 13.6 m/s
6. $3.9 \times 10^{12} \text{ Hz}$
7. $9.87 \times 10^{-3} \text{ J}$, at the extremes
8. 7.1 cm, PE also drops by half
9. Total energy is constant
10. 12.57 rad/s, 0.5 s, 9.51 cm
11. Circle of radius a

Chapter 6   Waves and their basic properties
1. $4.29 \times 10^{14} \text{ Hz}$ (red) to $7.5 \times 10^{14} \text{ Hz}$ (violet)
2. 49 N, 78.3 m/s
3. 2.04 km
4. Four times
5. Speed increases
6. 258.5 m/s
8. $2.26 \times 10^8 \text{ m/s}$
9. 0.3 wavelengths, 1.88 rad
10. Maximum, path difference 2 wavelengths
11. 336 m/s
12. 100 m/s, 20 N

Chapter 7   Aspects of optics
2. $50°$, $30°$, $100°$
3. 2.41, $1.24 \times 10^8 \text{ m/s}$
4. a) 30 cm, real and inverted, $m = 0.5$
   b) $(-)20$ cm, virtual and erect, $m = 2$
5. Rays emerge parallel
6. 0.75
7. b) since building subtends angle of $1.07°$
9. 25.17 cm, $\times 120$
10. 16.1 cm
11. $0.46°$
12. 1.03 m
13. For $\theta = 90°$, $n = 3.42$, hence third
14. 437 nm
15. $58.8°$

Chapter 8   Spectra and the electromagnetic spectrum
1. 4.09×10^{-19} J
3. 102.5 nm, ultra-violet
5. *e, c, b, a, d*
7. 3.09×10^{-10} m
8. *e, f*
9. 24 cm
10. 30°

Chapter 9   The nature of matter and the mechanical properties of solids
1. 1.04 mm
2. *a)* 9.98×10^7 N/m^2
    b) 4.99×10^{-4}
    c) 1.50 mm
    d) 0.059 J
3. 0.056°
4. 600 kN
5. 0.1 mm
6. 61.2 m/s
7. 1.58×10^{11} N/m^2
8. 0.17×10^{-6} m

Chapter 10   Heat, temperature and internal energy
2. 231 °C
3. −74.8 °C
4. $1.1985 R_0$, $1.3940 R_0$, 50.4 °C
5. 390 J/kg K
6. 411 J/kg K
7. 1120 J/kg K
8. 4.05 kJ/kg K, 0.6 W
9. 313 kJ/kg
10. 1107 kJ/kg, 2.75 W
11. 330 kJ/kg
12. 7.94:1
13. 3.20 W, 84.1 °C
14. 3750 W, 72.1 W
15. 48.4 W, approx. $5\frac{1}{2}$ hours, 0.97 MJ

Chapter 11   The gaseous state
1. 8.31 J/mol K
2. 2085 J/kg K
3. 779 mmHg
4. 3.47×10^{-3} per K, −288 K
5. 1.1×10^{-3} m^3
6. 8.56 mol, 0.3 mol
7. 735 mmHg, 712 mmHg, 203
8. 900 mmHg
9. 493 m/s
10. 576 m/s
11. 178.4 J, 14.3 K
12. 195.3 K

Chapter 12   The liquid state
1. 19 mmHg
2. 5.0 mm, 1.12×10^{-2} N
3. 1.023×10^5 Pa
4. 19.2 mm
5. 1.2 Poiseuille

Chapter 13   Electrostatics
3. 4.32×10^{-2} N
4. 3.60×10^7 N/C
    2.88×10^{-4} N
5. 4.86×10^{-3} J
6. 1.20 J
7. 1.5×10^4 J
8. 667 pC
9. 7.1×10^{-4} F
10. 1.95 nF
11. *a)* 2.1×10^{-6} F
    b) 6.3×10^{-6} C
    c) 2.1 V and 0.9 V
12. *a)* 42.9 V
    b) 128×10^{-3} C and 22×10^{-3} C
14. 1.28×10^{-10} J and 3.84×10^{-11} J

Chapter 14   Moving charges and circuits
1. 360 C
2. 1 Ω
3. 8R
4. 3:4
5. 4.08 J, 1.36 W
6. *a)* 240 V
    b) 960 W
    c) 3.46×10^6 J
7. 6.48 Ω
9. *a)* 0.143 A
    b) 81.8×10^{-3} W and 0.204 W
    c) 0.5 A and 0.2 A
    d) 1 W and 0.4 W
10. 1.5 Ω
11. 214 Ω, 1170 Ω or 420 Ω
12. 1.52 V

Chapter 15   Moving charges and magnetism
2. 3 T
4. 1.14×10^{-3} m
5. 6.28×10^{-3} T
    3.77×10^{-5} N
6. 2×10^{-6} T
7. 2.33 g
8. 0.01 N m
9. 3.77×10^{-3} T
11. *a)* 5×10^{-3} Ω shunt resistor required
    b) 9.985×10^3 Ω bobbin resistor required

Chapter 16   Electromagnetic induction
2.   0.1 V
3. 2.5×10^{-4} V
4. 1.78×10^{-3} H
6.   125 V
8.   1.78 V
9.   0.9 H

Chapter 17   Alternating currents
2.   8.49 A
3.   *a*) 339 V
     b) 678 V
     c) 0.12 A
     d) 0.17 A
     e) 17.3×10^3 J

Chapter 18   The electron and the photon
1.   64 kV
2. 10^{-14} kg
3.   124 eV
4. 2.5×10^{-13} m
5.   2.8 V
6. 4.8×10^{-19} J
7. 1.1×10^{15} Hz
8.   1.84 eV
10. 322 nm

Chapter 19   A model for the atom and emission spectra

1.   0.53 Å (53 pm)
2.   27 V
3.   434 nm
4.   122 nm
5.   *a*) 656, 486 and 434 nm
     b) 1875, 1282 and 1094 nm
6.   0.85 eV

Chapter 20   The nucleus
3.   0.17 *u*
4.   7.7 MeV

Chapter 21   The unstable nucleus—radioactivity
2. $_6\text{C}^{12}$, $_{83}\text{Bi}^{212}$, β
3. β
4. $_{94}\text{Pu}^{239}$
5.   3.4 min
6. $\frac{1}{4}$
7.   64.8 hours
8.   3.8 μg
9. 4.5×10^9 years

Index